Maria M. Sojka
A Heated Debate

AF765892

Philosophy

Maria M. Sojka studied philosophy and physics at Ruhr-Universität Bochum. Her research focuses on the philosophy of climate science, the philosophy of physics, the public understanding of science and feminist philosophy of science.

Maria M. Sojka

A Heated Debate

Meta-Theoretical Studies on Current Climate Research and Public Understanding of Science

[transcript]

Bibliographic information published by the Deutsche Nationalbibliothek
The Deutsche Nationalbibliothek lists this publication in the Deutsche Nationalbibliografie; detailed bibliographic data are available in the Internet at http://dnb.d-nb.de

Cover layout: Kordula Röckenhaus, Bielefeld
Printed by: Majuskel Medienproduktion GmbH, Wetzlar
https://doi.org/10.14361/9783839465806
Print-ISBN: 978-3-8376-6580-2
PDF-ISBN: 978-3-8394-6580-6
ISSN of series: 2702-900X
eISSN of series: 2702-9018

Printed on permanent acid-free text paper.

Contents

List of Abbreviations

AOGCM	Atmosphere-Ocean General Circulation Model
AR5	Fifth Assessment Report of the IPCC
AR6	Sixth Assessment Report of the IPCC
CMIP	Coupled Model Intercomparison Project
CTK	Collective Tacit Knowledge
DJ distinction	Distinction between the context of discovery and the context of justification
ECS	Equilibrium Climate Sensitivity
ESM	Earth System Model
GMST	General Mean Surface Temperature
IPCC	Intergovernmental Panel on Climate Change
MIP	Multi-model Intercomparison Project
MSU	Microwave Sounding Unit
NASA	National Aeronautics and Space Administration
NOAA	National Oceanic and Atmospheric Administration
NWP	Numerical Weather Prediction
RA	Robustness Analysis
RSS	Remote Sensing System
RTK	Relational Tacit Knowledge
SST	Sea Surface Temperature
STK	Somatic Tacit Knowledge
UAH	University of Alabama at Huntsville
UMd	University of Maryland
V&V	Verification and Validation

List of Figures

Acknowledgment

I would like to thank Helmut Pulte, who supervised the dissertation this book is the result of, for his guidance, advice and encouragement.

I would also like to thank Sebastian Weydner-Volkmann who kindly took over as second supervisor at the last minute.

Further, I would like to thank Jan Baedke, Benedikt Fait, Eva-Maria Fromm, Erdmann Görg, Anna Leuschner, Janelle Pötzsch and Maria Wargin for their advice and support.

While writing this dissertation I was also kindly allowed to spend some time at the Max-Planck Institute for Meteorology in Hamburg and I would like to thank Thorsten Mauritsen, Marco Giorgetta, Hauke Schmidt and everyone there who took the time to answer my many strange questions.

The research behind this book was also funded and supported by the Studienstiftung des Deutschen Volkes.

Last but by no means least a big thank you goes to Ariane Sojka, who read with inexhaustible patience countless revised versions of this book and Lasse Sojka, who helped out whenever my computer crashed or I had other IT problems. Annika Brünje and Sabrina Schenk, who were my comrade in arms during this time and my parents, who always fostered my love for science and philosophy.

1. Introduction

While scientists became increasingly confident over the second half of the 20th century[1] that anthropogenic climate change[2] is happening and will have severe effects on human life on earth, for a long time studies have shown that great parts of the public still perceived there to be no consensus about the topic within the scientific community.[3]

There are multiple reasons why the public perception that basic questions about the anthropogenic climate change are not yet settled was sustainable for such a long time. On the one hand, in the past the issue of climate change has often been reported in a similar fashion as political debates: as a debate with good arguments on both sides, neglecting that it is not a question of opinions but of facts (Edwards, 1999). Thereby, a distorted picture about the degree of general agreement and disagreement among climate scientists is created (Boykoff and Boykoff, 2004; Washington and Cook, 2011). On the other hand, institutions and individuals that have an interest in discrediting climate

1 In this context, it is often said that there is a 97 % consensus among climate scientists about climate change, referring to a study done by Cook et al. (2013). Although there is some disagreement about the exact number, several other studies have shown that the overwhelming majority of publications in climate science journals do not question the anthropogenic climate change (e.g., Oreskes, 2004).

2 That is, the climate is changing due to external, human-caused forcing. Generally, forcing refers to a change in the energy budget of the planet, which can happen by nature, e.g., radiation from the sun or erupting volcanoes but also CO_2 emissions produced by humans.

3 Studies from the *Pew Research Center* show that the perception of climate change as a major threat to humanity in many countries has increased since 2013. The survey across 23 countries displays an increase from a median of 56 % in 2013 to 67 % in 2018 (Pew Research Center, 2019). However, studies from the same institute also show that in 2014 only 57 % of adult U.S. citizens were under the impression that there was consensus among scientists that climate change is happening (Pew Research Center, 2015).

scientists' research results have been shown to have invested large sums of money into climate-change critical research projects (Oreskes and Conway, 2010). This has artificially amplified their voices and given the false impression that there is no consensus at all among scientists concerning anthropogenic climate change.

Multiple psychological factors also play a role (Van Lange et al., 2018). After all, many policies to mitigate climate change are inconvenient to most of us. This makes it much more compelling to disregard scientific evidence. Furthermore, the scale of the problem might seem so overwhelming that it appears like one's individual actions will not have an impact anyway. Another contributing factor is that (at least in what is often referred to as the 'Western world') we are only just starting to see the effects of climate change. Heat waves, droughts, wildfires and warm winters have brought climate change to the forefront of social and political debates, in the last few years.

However, I will argue in this book that there is another underlying structural problem concerning public understanding of science that so far has rarely been discussed: how climate-change deniers benefitted from certain ideals about how science does and should operate, which are widespread in the public understanding of science. However, over the last few decades work of philosophers of science has shown that science, specifically when dealing with high complexity in the target system, such as climate science, cannot and more importantly need not hold up to these ideals.

Even though actual scientific practice is often far from these ideals, one reason that these ideals are so prevalent is because they are also frequently perpetuated by scientists themselves when communicating with the public. Thus, the discrepancies between these ideals and the everyday work of scientists often only become visible when science is dragged into the spotlight due to its social relevance.

The objective of the book is twofold:

1. Three ideals about science, which are perpetuated in the public's understanding of science and play reoccurring roles in public controversies about climate science, are investigated. It will be shown by example of climate science why science cannot but also does not have to life up to these ideals.
2. Under the assumption that these ideals cannot be referred to in order to assess which scientific research results to have confidence in, the aim is also

to provide another route, specifically for outsiders to the scientific community, to determine when it is justified to be (at least) sceptical about the claimed expertise of specific individuals.

I deliberately use the plural *ideals* here. For one, to stress that what I will examine in the following by no means constitute an exhaustive summary of these kinds of beliefs. I will assert that the ideals examined in this book make reoccurring appearances in the context of different controversies about climate change. For another, because it seems prudent to assume that these ideals are not exclusive to the public discourse about the reliability of climate science but also return in different compositions in other instances where scientific research affects public life, I would like to make it clear that these different aspects do not make up one coherent and full ideal that will be realised in its entirety wherever science meets public scrutiny.

In the following, I will examine three of these ideals as they manifest themselves in the case of controversies about climate science. They concern the role of values in science, the relationship between data and theory, and the handling of uncertainty. While in the public debate about climate change these ideals take a prominent and reoccurring role, they can also all be interpreted as signs of a wish for more simplicity in science. Proponents of these ideals argue that:

1. scientists should not 'spoil' scientific research with their values, instead, produce unbiased irrefutable facts.
2. a scientific theory or model can be easily proven or refuted by a comparison to irrevocable observational or experimental data.
3. and science should give clear yes- and no-answers, so uncertainties only mean that scientists need to try harder.

These assumptions, however, greatly underestimate the complexity of modern science and the systems examined. One might argue that, at least to some extent, no science can entirely fulfil the ideals outlined above. But as I hope to show in the next few chapters, in a case study of climate science, this holds particularly for those sciences dealing with highly complex systems. These sciences hit what Johannes Lenhard calls the *complexity barrier* (2019, pp. 89–131), where the complexity of the system impedes access to full analytical understanding. Simulations, as they are done extensively in these sciences, are, he

argues, a way to circumvent these barriers, but they cannot overcome them.[4] The epistemic challenges coming from the complexity of the system make it distinctly evident that science functioning in such a simple and straightforward way as envisioned in these ideals is impossible. Considering this, the goal in the following is to show why science does not have to live up to these ideals.

To see how this will play out, let us first take a look at how these ideals are displayed in the public discussion of climate change: the first of these ideals of science states that science should be value-free (Chapter 3.1). The typical ideal of a 'man of science' is one who is fully detached from his personal beliefs on political or social issues in his work. While philosophers have discussed whether science can or even must uphold a separation between science and values since the early 20th century, the value-free ideal has been astonishingly persistent in the public perception of science as well as in science itself.

It seems reasonable to assume that this widespread perception of science as a fully value-free endeavour is also at the core of why it has been such as successful line of attack from climate-change denialists over the last few decades to frame climate change as a "hoax created by a conspiracy of supposedly greedy scientists, liberal politicians, and environmentalists" (Dunlap and Jacques, 2013, p. 713) and climate scientists having a personal "political agenda", as, for instance, former US President Donald Trump has done (BBC, 2018) or being "alarmists" (Medimorec and Pennycook, 2015) who overdramatise the situation (see also Brysse et al., 2013). Thus, it is implied that the scientists are influenced in an untoward way by their own 'leftist' values. Climate scientists also on occasion give the impression that they sometimes find themselves in the position where they feel like they have to defend themselves from these kinds of accusations (see e.g. Schmidt and Sherwood, 2015).[5]

---

4 Lenhard argues that the occurrence of the *complexity barriers* is not new to science. But historically, it was possible to overcome these without the help of computer simulations. As an example Lenhard cites the introduction of algebra in the 16th century (2019, p. 115).

5 For instance, at one of the high points of the attacks on climate scientists, Hans Joachim Schellnhuber, the then director of the *Potsdam Institute of Climate Impact Research* (PIK), pointed out in an interview that he drives a BMW, eats meat and is not a member of the *Green Party* (Evers et al., 2010). A somewhat paradoxical move from a scientist to call attention to the "value neutrality of his work by invoking certain values" (Leuschner, 2012a, p. 192). But Leuschner argues that Schellnhuber's declaration has to be inter-

The idea of the value-freeness of science in this context is often associated with the notion of *scientific objectivity*. Chapter 2.3 will show that the concept of scientific objectivity is not very well defined and has a variety of interpretations. Nevertheless, in the public understanding of science 'objective' science is often assumed to be value-free science. Thus, when climate science is attacked as being value-laden, it is commonly seen as not 'objective'. Only value-free science is considered to be objective and thereby good science.

As will be discussed in Chapter 3.1, philosophers of science have shown that science, specifically when its results have significant social and political implications, can never be guaranteed to be value-free. Nor does it have to be (Longino, 1990). Moreover, as Chapter 3.1.3 will show, the models used in climate science are commonly far too complex for climate scientists to consciously or inadvertently influence them in a specific direction furthering their personal social or political convictions.

The second ideal concerns the understanding of the relationship between experimentally acquired data und theory or, more specifically, theoretical models (Chapter 3.2). The common understanding of how the scientific process works is usually characterised in the following way: a scientist develops a theory or a hypothesis which then is tested by comparing it to data acquired by observation or experiment. Theories that cannot be proven by empirical data are to be disregarded immediately. In actual scientific practice, the relationship between the empirical and the theoretical is by no means that simple. At least since Thomas S. Kuhn published his hugely influential book *The Structure of Scientific Revolution* in 1962, there has been consensus among philosophers of science that scientists will usually need more than just one negative result of an experiment to overthrow a whole theoretical construct. Before Kuhn, philosophers such as Pierre Duhem (1906) and Willard Van Orman Quine (1951) also had argued that theories are underdetermined by empirical data. Furthermore, philosophers have established that observations are often (or even always) theory-laden. Norwood Russell Hanson (1958) is usually credited with having first clearly formulated the notion that our observations are influenced by theoretical background assumptions. In the context of highly complex computer models, philosophers have also argued that these models are not just theoretical constructs but also significantly data-laden (Edwards, 1999).

---

preted as "desperate reaction to the climate skeptics' standard argument" (Leuschner, 2012a, p. 192) of climate science being inappropriately value-laden.

Increases in the complexity of the systems explored also mean an expanding complexity in the data to be handled. In climate science data is collected on a global scale. As will be further discussed in Chapter 3.2.3, observational data – for example, satellite data – in climate science is also heavily model-filtered (Edwards, 1999). Meaning that the 'raw' data not only has to be assembled but usually has to be processed in terms of filtering and homogenisation and the like before it is of any use to climate scientists. Hence, contrary to what is sometimes assumed, it does not suffice to gather some data from a few thermometers and then compare them to the results of the models. The observations as well as the models are laden with uncertainties. Nevertheless, climate-change deniers have repeatedly and very successfully argued that the disagreement between data sets and models does unequivocally show the models' failing (Lloyd, 2012).

A striking example of this is the infamous *Climategate scandal*, during which private emails between climate scientists from *Climatic Research Unit* at the University of East Anglia were leaked and the scientists, subsequently, accused of illicit data manipulations. In the emails the scientists were discussing using a "trick" to treat their data in order "to hide the decline" in tree-ring proxy data (Jones, 1999). This turned out to be a normal and well-established practice in the climate science community to counterbalance the so-called *divergence problem*.[6] Even though the incident was thoroughly investigated and the scientists later acquitted of any wrongdoing by several independent investigations, the media coverage gave the impression that there had been serious misbehaviour by the scientists in question (see also Leuschner, 2012b, pp. 39–47; Oxburgh et al., 2010).

An oversimplified view of the relationship between theory and data can effect actual climate policy. At the turn of the century climate-change sceptics have argued – quite successfully at the time even in front of the US congress – that an apparent discrepancy between models' predictions and data from satellite and weather balloons would show unequivocally that the climate models have been wrong and that, therefore, the models' prediction of temperature

---

6 To be more specific, the 'trick' here refers to a way of homogenising data so as to deal with the widely acknowledged problem of a "dramatic change in the sensitivity of hemispheric tree-growth to temperature forcing" (Briffa et al., 1998, p. 65) observed in the second half of the 20th century, that "if not recognized and accounted for, could lead to erroneous inferences about past and future climate changes" (Briffa et al., 1998, p. 66).

rise due to climate change have been vastly overestimated and no mitigating actions have to be taken (Edwards, 2010, pp. 413–418; Lloyd, 2012). However, instead of discarding the models many climate scientists questioned the adequacy of the particular data set. As Lloyd has pointed out with respect to this controversy, climate data ought not to be treated as "windows on the world, as reflections of reality, without any art, theory, or construction interfering with that reflection" (Lloyd, 2012, p. 392).

The third ideal that will be discussed relates to the expectation that science provides predictions with a clear yes-or-no-answer (Chapter 3.3). According to this view of science, uncertainties are a sign of premature science where the scientists have not done their job properly. The ability to make predictions has become a hallmark of modern science. Specifically, a scientific discipline which research has significant implications for society is expected by policy-makers and the public at large to give distinct binary answers. So, when scientists voice uncertainty, it is often interpreted as no knowledge, not as knowledge to a certain degree.[7] This, of course, does not take into account that all scientific research is tainted by some degrees of uncertainty. Scientists are well aware that their work always has a preliminary character and might very well be overturned someday. Thus, this ideal is in stark contrast to what science actually can achieve.

In the public debate about climate change, this has manifested itself in the way climate-change sceptics have argued against taking mitigating measures. Uncertainties to some degree are presented as a sign that there is no evidence at all that anthropogenic climate change is happening, with the implication that there is no reason to act. The argument that the science on climate change is not yet settled (Howe, 2014) discounts, on the one hand, the high complexity of the climate system and, on the other hand, negates that, concerning the basic questions – for example, how increasing the amount of CO_2 in the atmosphere will lead to a global mean temperature rise – there is wide reaching consensus in the climate science community.

In situations that require urgent action, waiting for 'more certainty' comes, of course, at a certain cost. All in all, the question of how much certainty is

7 It has to be acknowledged here that in certain forms probability statements in science are generally accepted, even by a public audience. One might think, for instance, of quantum mechanics, where uncertainty is inherent to the non-deterministic quantum mechanical system. With a deterministic system, like the climate system, the expectation is often still that scientists deliver clear and precise research results.

required and at what stage to act is at best one of cost-benefit analyses. In the worst case, by the time the certainty is considered 'sufficient', it might be too late to act.

Thus, an insufficient public understanding of science has serious implication that go beyond the epistemic. In a case like climate change that has clear public and social significance, a lack of public understanding of how science actually works makes it easy for certain interest groups to undermine climate scientists when they raise legitimate concern about the threat of climate change. Instead the scientists are declared 'alarmists': that is, they are accused of "over-interpreting or overreacting to evidence of human impacts on the climate system" (Brysse et al., 2013). This makes it so dangerous to uphold these ideals. In doing so, science itself undermines its relevance to the public discourse:

> The danger is that holding science up to the wrong standard will diminish the value of what science discovers about nature, and could create an environment in which science is no longer consulted to inform policy. "The scientists can't give us a definitive answer, so why should we listen to them?" (Mitchell, 2009, pp. 118–119)

Still, from within the scientific community these simplistic ideals are quite often kept alive as a kind of 'useful fairy tale'. They underline the importance and infallibility of science. And after all, they "do not do any harm to the scientists unless policymaker start to believe that science is really so simple" (Collins, 2014, p. 24). This, however, can turn into a problem in those instances where the work of scientists is watched more closely than usual by the public.

Nevertheless, if we throw out these ideals, the question remains how can one then – specifically as outsiders to the scientific community – tell if the scientists and their work are trustworthy. In the following I will offer a different approach to this problem. Instead of resorting to either the virtues of the scientists or some distinct methodological approach that scientists follow, I will highlight the relevance of specialist tacit knowledge in science in general, which gains in relevance in the context of increasing complexity in science. Although tacit knowledge is at odds with the depiction of science in the aforementioned ideals, as it is commonly seen as something subjective and personal, I will argue that acknowledging this role of tacit knowledge, in fact, opens up a chance, even for outsiders to the scientific community, to assess at least to some degree whether or not to trust the work of the scientists. The argument, in short, is the following: the relevance of tacit knowledge also

highlights the necessity of experience for expertise. Following Collins and Evans (2009), I will claim that this leads to a concept of scientific expertise that puts the experience gained by scientists while working in their specialist field front and centre. This definition of expertise is specifically useful in a situation where science is under distinct public scrutiny and researchers are confronted with criticism from individuals who are presented to the public as apparent 'experts' but who have never actually worked in the specific subject in question. As will be discussed further in Chapter 4, it can be shown that some of the most prominent 'experts' that climate sceptics have referred to in order to undermine climate science actually have never done any specific research in the field of climate science (Oreskes and Conway, 2010, p. 8). Further, I will argue that, if tacit knowledge plays such an important role in science, then the place where it is acquired, namely, the scientific institutions, has some specific relevance. The structures of these institutions luckily are much more accessible to an investigations, even for outsiders, than the models or other methods used by the scientists.

In regard to the structure of the following chapters, I will start with some preliminary remarks in the next one. The aim here is to introduce some recurring themes that will be relevant in the then following discussion of the above mentioned ideals about science: the epistemic challenges of highly complex systems, the distinction between the context of discovery and the context of justification and scientific objectivity. The *complexity* of the climate system and the resulting *additional epistemic challenges* are what makes the failure of these ideals so apparent. The distinction between *context of discovery and context of justification* is a constitutive element to two of these ideals (see Chapter 3.1 and 3.2). In the context of science the term *objectivity* has become almost synonymous with 'good science' and, therefore, different interpretations of *scientific objectivity* will be significant when discussing certain idealised representations of science.

Chapter 3 then will focus on an in-depth discussion of three prominent ideals about scientific methods and objectives in relation to climate science: value-freeness, a clear separation between theoretical and empirical work, and the claim that science has to provide clear, binary answers. Each subchapter in Chapter 3 corresponds to one of these ideals and is structured in a similar way. I will start with a short historical introduction how the specific ideal came into being. These subchapters are not supposed to recap the full history of these ideals. The point is rather to trace where these ideals have come from and how they

have risen to prominence. After that a small subchapter will follow introducing one or two central philosophical concepts or issues that are of relevance in the context of the conflict of the ideals with scientific practice. The third and central part of every one of these chapters then is a discussion about how these ideals cannot be fulfilled in the context of climate science, but also why they do not have to. In the conclusion of Chapter 3 I will return to the concepts discussed in Chapter 2. A direct examination how they fare in the context of climate science will point to a way forward how to circumvent the problem that Chapter 3 leaves us with; that is, that these three ideals cannot be resorted to in order to assess the quality of scientific research. It will be shown that the increasing complexity gives tacit knowledge a more significant and more visible role in science.

Chapter 4 will discuss tacit knowledge in more detail. It will be argued that the epistemic challenges of complex climate simulations, particularly the difficulties of reaching "analytical understanding" (Lenhard and Winsberg, 2010), makes the reliance on tacit knowledge in science particular visible, but also grounds scientific research in the institutions and communities where this tacit knowledge is acquired and created. Further, a concept of expertise derived from tacit knowledge is introduced as an alternative to the failed ideals examined in Chapter 3.

Chapter 5 forms the conclusion and provides an outlook what all of this means for science, philosophy and society.

2. Some preliminary remarks

Before examining three popular ideals about how science operates and what constitutes good science and how they fail in the context of climate science, I will first introduce some terms and concepts that will be of relevance throughout this discussion: the epistemic challenges of complex systems, the distinction of context of discovery and context of justification in philosophy of science, and the concept of scientific objectivity. These are not ideals in their own right but rather form 'recurring themes' and are presented here separately from the aforementioned ideals because they are essential to the following discussion in two ways: on the one hand, they either play a constitutive role in the development of these ideals or show how the ideals fall short in the context of modern sciences. On the other hand, a closer investigation of how these motifs 'behave' specifically in the context of climate science will give an indication how the gap left by these ideals can be circumvented. At the end of the next chapter (Chapter 3.4) it will be shown that one element these 'recurring themes' have in common when viewed through the lens of climate science is that they highlight the relevance of the experience or skill that scientists develop through their work. Following from this, a concept of expertise, based on this kind of experience, is discussed in Chapter 4. There it is argued that this concept can – in some public debates about the trustworthiness and reliability of specific scientific research – function as a substitute for the ideals.

2.1 Epistemic challenges of highly complex systems

It seems to be a natural tendency of science to investigate increasingly more complex systems for two interconnected reasons. On the one hand, scientists turn their attention to ever more complicated questions. One way to do so is to examine continuously more complex systems. These, on the other hand, be-

come also more and more accessible to scientific research as the technology advances in a way that creates new instruments to explore these complex systems. Specifically, computer and computer simulations have been significant in tackling complexity in science (Lenhard, 2019).

Although the term is not sharply defined, complexity in science is often loosely understood "as a consequence of numerous independent interacting parts" (Strevens, 2016, p. 696; Weaver, 1948). A complex system usually refers to non-reducible systems with certain characteristics such as non-linearity, emergence, interactivity and path-dependency. The climate system is a complex system par excellence and fulfils, as the next few chapters will show, all these requirements. To assess the epistemic challenges of climate science, the complexity of the system in question is essential. However, I should note here that in the next few chapters I will not continue to assess to what extent exactly the climate system is complex or what particularly defines such a system. Instead when it comes to understanding why climate science cannot fulfil specific expectations about how science is supposed to function, a coarser definition of complexity will be sufficient. The relevant question here is not so much what exactly defines complexity but what follows epistemically from the fact that the climate system – and the computer simulations used to explore it – shows a broad variety of features of complex systems. More specifically, what is particularly significant here is that the complexity has consequences for the question to what extent understanding (of the climate systems and the models) can be achieved. What does that mean? Let's take a look at the particular epistemic challenges of climate modelling.

At the core of modern global climate models, so called Earth System Models (ESM), are a number of basic partial differential equations based on well-established principles and laws of physics, such as Newton's second law, thermodynamics and Navier-Stokes equations.[1] Historically, these mathematical descriptions of the dynamics of the climate system arise out of what are today referred to as the *primitive equations*, first developed by Vilhelm Bjerknes in 1904.[2] However, because these equations cannot be solved analytically, climate

1 For a more elaborated history of the development of climate models see Weart (2010); Gramelsberger (2011) and especially Edwards (2010).

2 To be more precise, Bjerknes proposed that the dynamic of the weather system could be described by these seven equations:
"1. The three hydrodynamic equations of motion. These are differential relations among the three velocity components, density and air pressure.

scientists have resorted to computer simulations. This is done by discretising these analytically unsolvable equations to calculate the state of the system step by step in specific intervals of time. In the case of global climate models, this means that the globe gets covered by a virtual three-dimensional grid (see Figure 1).[3] Information about climate variables such as temperature, pressure, humidity and wind are then determined for every grid cell at discrete time steps and is shared with neighbouring cells. Although this is often referred to as a *numerical solution*, the transformation of the analytical equations is only an approximation and the source of some uncertainties in climate modelling (see Chapter 3.3.3).

Figure 1: Discretisation grid for a climate model

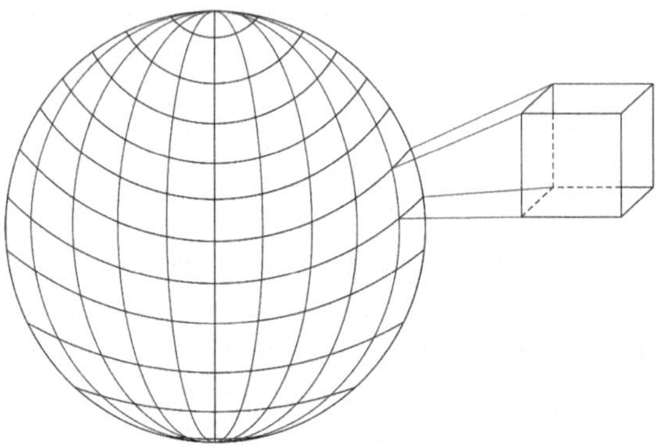

2. The continuity equation, which expresses the principle of the conservation of mass during motion. This equation is also a differential relation, namely between the velocity components and the density.
3. The equation of state for the atmosphere, which is a finite relation among density, air pressure, temperature, and humidity of a given air mass.
4. The two fundamental laws of thermodynamics, which allow us to write two differential relations that specify how energy and entropy of any air mass change in a change of state" (Bjerknes, [1904] 2009, p. 664).

3 Typical depictions of climate models show cube-shaped grid cells (like in Figure 1). Those have the disadvantage that they differ in size in a spherical system (i.e., the closer to the pole, the smaller they get). To tackle this inconsistency in grid size many contemporary climate models now have other grid shapes, e.g., icosahedral grids (Zängl et al., 2015).

Processes deemed too complex and/or (more importantly) taking place on a scale that is below the grid size are integrated into the models in form of parametrisation (McFarlane, 2011).[4] Processes, such as cloud formation, radiation, atmospheric convection but also tree growth, are happening on a scale that is too small and/or are not well enough understood to express in terms of physical laws to be represented in a resolved way at the grid-scale level. Nonetheless, these processes are considered to be too important to the climate to be just left out of the model. So to include these kinds of processes, scientists developed parametrisations that serve as substitutes for sub-grid level processes in the form of functions of above-grid-level variables.

Parametrisations are sometimes called being "semi-empirical" (for example Edwards, 1999, p. 449; Parker, 2018) as well as having been noted to include "non-physical" elements (for example Winsberg, 2018, p. 48). On the one hand, parametrisations are usually only partially based on insight into and physical description of the climate system but also on approximations gained from observations. On the other hand, usually some parameters in parametrisations are not physically constrained and do not have a direct equivalent in the actual target system. Thus, they are not set to represent a real-world 'true' value but a 'best' value. This makes parametrisations, both in the necessity for them and the way they are constructed, "artifacts of the computation scheme" (Winsberg, 2018, p. 49). Creating parametrisation schemes can be a highly complex process and there are frequently different options of how to parametrise the same process. This makes parametrisation a critical source for uncertainties in climate modelling (see Chapter 3.3.3).

The development of global climate models is an intricate and time-consuming task. A typical climate model of the scale of an *Earth System Model* (ESM), which are currently considered the "state-of-the-art" climate models, consists of hundreds of thousands lines of code, written and further developed by several generations of scientists – often over more than a decade. Because of this, these kinds of models are usually not built from scratch but incorporate bits and pieces from previous generations of models (Knutti et al., 2013). It is also not uncommon for scientists to 'borrow' parts of or even whole model components from other institutions and modify it so that it fits with their model

4 Climate scientists commonly differentiate between the physics and the dynamics of an atmospheric model. The former refers to those processes that have to be parametrised and the latter to the dynamical processes that can be described in a resolved form.

when they lacks that specific expertise (Alexander and Easterbrook, 2015, pp. 1227–1228).

Structurally, a climate model is constructed out of several different components. A typical ESM consists of, for instance, an atmosphere, an ocean, a sea ice, a land and vegetation component.[5] The exchange of relevant variables at the border regions of the different model components is facilitated via a so-called *coupler*. A coupler compensates for the difference in resolution of different model parts and supports the exchange of information between the various model components.

Climate models are "highly modular" (Winsberg, 2018, p. 50). There are a number of different ways the coupling of the different model parts can be constructed. For example, a study by Alexander and Easterbrook (2015) shows that models coming from American modelling centres often have a "star-shaped" structure, where every component is connected directly through a coupler to the other components. Models from European climate modelling institutes tend more towards a "two-side" structure, where only the atmosphere and ocean component are directly connected to the coupler. The land, sea ice and other possible elements are integrated into either the ocean and atmosphere component requiring that the resolution of minor components is the same as the atmosphere or ocean model, respectively (Alexander and Easterbrook, 2015, p. 1225). In a similar fashion there are usually a variety of ways in which specific climate processes are integrated into a model. Depending on the resolution of the model, particular processes need to be parametrised or can be integrated into the model in a resolved way. Further, because of limited computing power, time-constraints and previous modelling decisions not all processes can be represented equally well in a model. How a model component

---

5 Alexander and Easterbrook note that there are certain analogies between how the expertise in climate modelling facilities, the computer model and nature itself is structured: "The boundaries between components in an Earth system model represent both natural boundaries in the physical world (e.g., the ocean surface) and divisions between communities of expertise (e.g., ocean science vs. atmospheric physics). The model architecture must facilitate simulation of physical processes that cross these boundaries (e.g., heat transport) as well as support collaboration between knowledge communities within the work practices of model development (e.g., to study climate feedbacks)" (Alexander and Easterbrook, 2015, p. 1224). However, it should be mentioned that there are, of course, also limitations to these similarities and one should not assume that nature 'functions' in the same way as the models (specifically with respect to parametrisations).

is structured, whether a process is integrated directly or through a parametrisation, which processes and variables get preferential treatment leaves many different options of how to build a climate model (Parker, 2018). In practice, how these decisions are made often depends on different modelling traditions and trade-off considerations (see Chapter 3.1.3). That is, that there is no such thing as a 'universally agreed upon construction manual' for climate models.

One might now think that, because ESMs are composed of many different parts, this means that, even though the models as a whole are complex and large structures, the quality of the models can still be assessed relatively easily by testing the different model parts separately. However, there is a high interdependency and exchange between various elements of the models. In practice, the components of the model are, of course, also tested extensively and calibrated on their own but will inevitably perform differently when put together (Hourdin et al., 2017, p. 591). Lenhard and Winsberg have called this the "fuzzy modularity" of climate models:

> In sum, climate models are made up of a variety of modules and sub-models. [...] And it is the interaction of these components that brings about the overall observable dynamics in simulation runs. The results of these modules are not first gathered independently and then only after that synthesized. Rather, data are continuously exchanged between all modules during the runtime of the simulation. The overall dynamics of one global climate model is the complex result of the interaction of the modules—not the interaction of the results of the modules. For this reason, we like to modify the word "modularity" with the warning flag 'fuzzy': due to interactivity, modularity does not break down a complex system into separately manageable pieces. (Lenhard and Winsberg, 2010, p. 256)

The high interdependency and continual exchange between the different parts of the model can also give rise to compensating effects to the extent that some feature of one model part will interfere with another element of the model in such a way that it makes up for some particular shortcomings of the individual model component.

Further, climate modelling requires tuning. Tuning is a process at the end of a model- or submodel-construction cycle, where a few parameters (for example, parameters concerning cloud or surface albedo properties) are adjusted so that the model behaves in a way scientists consider to be realistic (Mauritsen et al., 2012). Although there is some common consensus about some general aspects of the tuning process and goals, how a specific model is tuned depends

considerably on the objectives and preferences of the modelling group in question (Chen et al., 2021, pp. 217–218).[6]

As will be discussed further in Chapter 3.1.3, tuning is always a question of trade-off, meaning that one cannot tune a 'perfect' model in respect to every variable. Climate scientists have also voiced concern that tuning models for 20th century warming may inadvertently lead to and mask compensating errors (e.g., Mauritsen et al., 2012): that is, that the improved model performance is rooted in hidden structural problems.

Lenhard and Winsberg have famously argued that all this makes climate models "analytical impenetrable in the sense that we have been unable, and are likely to continue to be unable, to attribute the various sources of their successes and failures to their internal modeling assumptions" (2010, p. 261). With reference to William Wimsatt's concept of "generative entrenchment" (2007, p. 133), they argue that the models' "layered history" (Parker, 2018) intricately influences the performance of the model. On the one hand, many aspects of climate-model development are not fully epistemically constrained. That means that there is usually more than one option how to integrate particular features of the climate system into a model. On the other hand, what kind of steps can or will be taken next in the continual development of a model is constricted by previously made modelling decisions. This *path-dependency* of the models means that climate modellers are not infrequently limited in their modelling choices by deliberations of previous generations of scientists and they will, in turn, further constrain the options and strategies available in the future. Further, there are always trade-offs to be made and, because there is a limit to computing power and time available, not all relevant climate processes can be represented equally well within the model (see Chapter 3.1.3).

This goes hand in hand with what Lenhard and Winsberg, with reference to Andy Clark (1987), call *kludging*. Kludging describes the piecemeal construction of complex computer programs that is effective but, nevertheless, "botched together" (Clark, 1987, p. 278) to the extent that as it does not follow a coherent construction plan. This can contribute to a lack of or reduction in analytical

---

6 A common target of tuning, for example, is that the mean equilibrium temperature the models display is in agreement with observations. A frequently used strategy to achieve this is to regulate the top-of-the-atmosphere energy balance through adjusting, e.g., cloud parameter. The adjustment of these parameters is important as global warming essentially is an energy imbalance (i.e., a mismatch between the incoming and outgoing energy) of the system (Hourdin et al., 2017).

understanding of the program. Considering what we have learnt so far about the specific characteristics of climate model developing – multiple generations continually adjusting the project, often driven by pragmatic considerations – it is easy to see why kludging would both occur here and further affect the access to analytical understanding.

There is some disagreement among philosophers to what extent these limitations in analytical understanding are 'here to stay' or can be overcome (at some point). Lenhard and Winsberg (2010) argue that climate modelling is affected by a strong form of *confirmation holism* often preventing a precise attribution of the source of error in a model. Parker (2018), however, notes that there has been some progress in gaining analytical understanding in recent years (e.g., through finding so-called *emergent constraints* (see Chapter 3.3.3.4)). It is, nevertheless, a difficult and laborious undertaking.

This gives us a first glimpse of the epistemic challenges of the high complexity of the climate system and the models climate scientists employ to tackle this complexity. It also indicates why the failure of certain ideals of science becomes so apparent in the wake of the difficulties described above in gaining analytical understanding. All these aspects will be discussed further in the following chapters.

It should be pointed out here that all sorts of different types of models are used in climate science. Although I will primarily focus on ESMs (*Earth System Models*), which are the state-of-the-art global climate models, and their predecessors AOGCM (*Atmosphere-Ocean General Circulation Models*),[7] climate scientists also make use of regional models to estimate climate development on a local level. Global models of various complexities lower than that of an ESM are also frequently applied in those cases where a reduced demand in computing power is advantageous (Chen et al., 2021, pp. 218–219). Many of the issues and philosophical challenges to be discussed in the next chapter also apply to these models. To understand the impact of climate change and options of mitigation, scientists also employ other types of models such as *Integrated Assessment Models* (IAM), which will also not be discussed here.

7 Compared to AOGCMs, ESMs also include biochemical processes (Chen et al., 2021, p. 181).

2.2 Discovery and justification: the DJ distinction

When Hans Reichenbach published his landmark book *Experience and Prediction* in 1938, his primary intention might have been to introduce himself and his brand of philosophy to the American philosophy of science community after he had to emigrate from Germany (Howard, 2006, p. 7). But the first chapter also spelt out a concept that would impact the (self-)perception of philosophy of science for the rest of the century: the distinction between the context of discovery and the context of justification (usually abbreviated to *DJ distinction*). The DJ distinction significantly reduces the scope of philosophy of science and, thereby, has had subsequently a (sometimes somewhat concealed, sometimes more obvious) influence on many discussions and controversies in philosophy of science. It should, therefore, not be surprising that we will see this distinction popping up in several places throughout the discussion of specific popular ideals about how science ought to operate in Chapter 3. The reason why I introduce this concept and some modern interpretations here separately from any distinct ideal is twofold: first of all, the DJ distinction is constitutive not just for one but two of these ideals. Secondly, as we will also see in the next few chapters, in the context of complex computer simulation, as they are used in climate science, even a "lean version" (Hoyningen-Huene, 2006) of the DJ distinction cannot be upheld as it is difficult to fully separate the evaluation of these simulations from their history (see Chapter 3.4.2).

But before we turn our attention to any modern interpretation of the DJ-distinction, what was Reichenbach's original reasoning? His overall claim is that there are three tasks epistemology has to tackle: the *descriptive task*, the *critical task* and the *advisory task*.

The first task of epistemology is "giving a description of knowledge as it really is" (Reichenbach, 1938, p. 3). This, however, does not mean that the epistemologist should be concerned with describing all and any of the thoughts the scientists actually had before coming to a particular conclusion. That sort of reconstruction of a scientific thought process falls, argues Reichenbach, into the realm of psychology. Instead, the job of the epistemologist is to perform a "rational reconstruction" (Reichenbach, 1938, p. 5):

> What epistemology intends is to construct thinking processes in a way in which they ought to occur if they are to be ranged in a consistent system; or to construct justifiable sets of operations which can be intercalated between the starting-point and the issue of thought processes, replacing the real

> intermediate links. Epistemology thus considers a logical substitute rather than real processes. (Reichenbach, 1938, p. 5)

Hence, what Reichenbach has in mind is a logical reconstruction of an ideal thought process somewhat comparable to the reconstructed thought processes which scientists themselves publish in scientific journals to communicate their findings to their peers.[8]

The second task of epistemology is the *critical task*. It overlaps in some ways with the *descriptive task* but must, as Reichenbach insists, be viewed separately, for its central objective is not just to describe but to criticise "the system of knowledge [...] in respect of its validity and its reliability" (Reichenbach, 1938, p. 7). Besides examining the logical basis of science, another main function of the critical task is to point out "volitional decisions" (Reichenbach, 1938, p. 9). Reichenbach acknowledges that the scientific process includes many instances in which the next step cannot be determined by logical deliberations alone. Instead, scientists routinely have to make methodological decisions between two or more equally good options. Detecting and disclosing these "volitional decisions" is "one of the most important tasks of epistemology" (1938, p. 9). This includes the specification of conventions (for example, measuring units) and "volitional bifurcations" (Reichenbach, 1938, p. 10), compared to conventions these are decisions which do not result in equivalent systems.

The third task is the *advisory task*. Unlike what one might surmise from the name, the advisory role of epistemology must, according to Reichenbach, be curtailed to the bare minimum. That means the epistemologist must refrain from giving direct advice which decisions to take. Instead, Reichenbach argues that the only appropriate part for philosophers in the decision-making process in science is to point out different available options:

> We may therefore reduce the advisory task of epistemology to its critical task by using the following systematic procedure: we renounce making a proposal but instead construe a list of all possible decisions, each one accompanied by its entailed decisions. So we leave the choice to our reader af-

8 Reichenbach, however, stresses that common scientific writing is not precise enough for the kind of logical reconstruction epistemologists should aspire to: "For scientific language, being destined like the language of daily life for practical purposes, contains so many abbreviations and silently tolerated inexactitudes that a logician will never be fully content with the form of scientific publications" (Reichenbach, 1938, p. 7).

> ter showing him all factual connections to which he is bound. (Reichenbach, 1938, p. 14)

Reichenbach emphasises that the role of philosophers here is at best to make a "proposal" by calling attention to the advantages and disadvantages of a decision, not a "determination of truth-character" (Reichenbach, 1938, p. 13). One particular concern for philosophers in this context, Reichenbach points out, are what he calls "entailed decisions" (Reichenbach, 1938, pp. 13–16), that is, tracking and identifying the consequences of decisions. In this situation, the philosopher is, according to Reichenbach, in the position to show how specific disputed decisions logically follow from already well-established ones.

Reichenbach, thus, here sets clear boundaries for the scope of philosophy of science. Its work should be restricted to the reconstruction and evaluation of scientific arguments from a logical point of view. The advisory role of epistemology is limited to pointing out different options and does not extend to an interference in the actual decision-making processes.

Setting the scope of philosophy of science in terms of a distinction between the realm of discovery and the realm of justification quickly became prevalent in philosophy of science for some time. Beginning in the 1960s and at the start of the decline of the dominance of logical empiricism in philosophy of science, some philosophers began to voice criticism. They argued against the omission of the dimension of discovery and history from philosophy of science (Schickore and Steinle, 2006a). Instead, they argued for a "logic of discovery" (Nickles, 1980), for including historical analysis and for returning scientific practice and experimentation back into the philosophical limelight (see Chapter 3.2.1).

Paul Hoyningen-Huene (2006) comes to the conclusion that some of the resistance against the DJ distinction can be traced back to some confusion that arose from the fact that by the mid-century there were several different versions of the DJ distinction at play often muddled together. According to Hoyningen-Huene, this led to a situation where in respect to the DJ distinction it was "not clear what exactly is stated by its defenders and what exactly is attacked by its critics. Eventually, all parties, growing frustrated, turned away from the discussion" (Hoyningen-Huene, 2006, p. 119). As the DJ distinction will be a recurring topic in the next chapter, it is worthwhile to take a closer look at some of the different versions of the DJ distinction that Hoyningen-Huene identifies (2006, pp. 120–123). Historically, he claims there are the

following five different variations of the DJ distinction in the literature of the 1960/70s:

1. two temporally distinct processes
2. the historical discovery process versus the specific justification methods
3. an empirical versus logical process
4. a disciplinary distinction
5. a differentiation in respect to different questions asked

First, there is the temporal distinction. Here discovery and justification are seen as two processes taking place one after the other. Initially, there is a discovery (here the term can be stretched to include inventions), which is followed by a justification process. This definition of the DJ distinction does not hold up to actual scientific practice. The second definition focuses on a distinction between discovery processes and justification methods. That is, there is a "contrast between the factual historical process and methods, considerations, procedures, etc. that are relevant to justify or to test knowledge claims" (Hoyningen-Huene, 2006, p. 121). Hoyningen-Huene argues that this can be either interpreted historically, running into the same problem as the first definition or normatively; meaning that "historical processes (of discovery) are described, whereas claims of justification or testing are normatively evaluated" (2006, p. 122). One can specify this version further by defining discovery as an empirical process and justification as a logical process. Following this distinction, one might also attribute different disciplines to the two categories. History, science and psychology of science are considered to be methodologically empirical, whereas philosophy of science is methodologically logical. The fifth version of the DJ distinction that Hoyningen-Huene identifies in the literature is that of two different and distinct questions being asked "such as 'What has happened historically during this discovery?' versus 'Can a statement be justified? Is it testable?'" (2006, p. 123). Hoyningen-Huene argues that specifically versions 1–4 (historically) are often merged into one,[9] leading to a logical em-

9 The implication of this, Hoyningen-Huene asserts, is that "*a rational disagreement about justification is conceptually impossible*" (2006, p. 124). It is assumed that the sphere of discovery can only be subjected to empirical investigations. Thus, there is no place for philosophy and there is no such thing as a "logic of discovery". The mixing of different versions of the DJ distinction also leads to the assumption that the only form of justification is logical and it is the 'business' of philosophy of science to evaluate it, Hoyningen-Huene argues. That means that any conflict on grounds of questions of justification

piricists' understanding of what constitutes philosophy of science (2006, pp. 123–124).

Hoyningen-Huene himself, subsequently, argues for what he calls a "lean" version of the DJ distinction that includes elements of version 2 and 5 (2006, pp. 128–130). The general idea is that there is a difference between a *descriptive* and a *normative perspective*:

> From the descriptive perspective, I am interested in facts that have happened, and their description. Among these facts may be, among other things, epistemic claims that were put forward in the history of science, that I may wish to describe. From the normative or evaluative perspective, I am interested in an evaluation of particular claims. In our case, epistemic claims, for instance for truth, or reproducibility, or intersubjective acceptability, or plausibility, and the like are pertinent. Epistemic norms (in contrast to, say, moral or aesthetic norms) govern this evaluation. By using epistemic norms we can evaluate particular epistemic claims according to their being justified or not. (Hoyningen-Huene, 2006, p. 128)

Hoyningen-Huene argues that what makes this version "lean" is that it merely distinguishes between two different perspectives. Thereby, one does not have to make any additional assumptions, such as that there cannot be any overlap between the two spheres both in a categorical and a procedural sense (see also Chapter 3.4.2).

The context distinction, as envisioned by Reichenbach and, subsequently, the logical empiricists, limiting specifically the scope of philosophy of science to a logic of justification is now seen as out-dated by the vast majority of philosophers of science. But the aftermath is still felt, argue Schickore and Steinle:

> In recent years, philosophers have rarely directly addressed, let alone attacked the distinction. But this does not mean that the distinction has been rendered irrelevant or that it has been successfully refuted. On the contrary, the legacy of earlier advocates of the distinction is still effective, and the distinction continues to delineate the scope [*sic*] philosophy of science. (Schickore and Steinle, 2006a, p. ix)

---

can only emerge with respect to either an error or differences in conventions. However, as differences in conventions are not considered to be "epistemically substantial disagreements", the analysis of potential errors in the logical justification "is a one-person-game" (Hoyningen-Huene, 2006, p. 124).

Schickore and Steinle see both the separation and lack of exchange between philosophy of science, history of science and science studies as well as the lack of interest from the analytically orientated philosophy of science, only rarely integrating knowledge from the other two disciplines, as a lasting sign of this (Schickore and Steinle, 2006a, pp. ix–x). And, as already mentioned, it also had a prolonged influence on how philosophers, scientists[10] and the public have idealised or still idealise the inner workings of science. For this reason, the DJ distinction will also be crucial for understanding the history of two of the three ideals of how science does and should operate discussed in the next chapter. On the one hand, the notion that science should disregard sociological and psychological aspects of science was a constitutive element to the ideal of value-free science (see Chapter 3.1). On the other hand, the DJ distinction was also a significant contributing factor to the neglect of the experimental part of science by philosophy of science during much of the 20th century, as the experiment is traditionally seen as mostly an element of the discovery side of science that Reichenbach has declared to be of no philosophical interest (see Chapter 3.2).

In the context of climate modelling, it will also become apparent that the DJ distinction is problematic even in the weaker form that Hoyningen-Huene proposes. As will be discussed at the end of the next chapter (Chapter 3.4.2), in the context of climate science, it is no longer possible to fully separate the evaluation and justification of models as well as the techniques employed in their construction and evaluation from their own history. This highlights the relevance of the experience climate scientist develop in working with the models, which will explain why a conception of expertise rooted in this experience can be a successful way for those who are outside of the scientific process to assess when to be sceptical about claims made by apparent scientific 'experts'.

---

10 Interestingly, Schickore and Steinle argue that the DJ distinction is still very much alive in science itself: "Remarkably, today the distinction is most explicitly discussed in the sciences themselves. In methodological introductions of science textbooks, it shapes the regulations for scientific research. These textbooks employ a particular version of the distinction, namely the context distinction temporally understood in combination with the hypothetico-deductive (H-D) model of scientific research" (Schickore and Steinle, 2006a, p. ix). Schickore and Steinle however also note that, when scientists perceive this to be the actual view that philosophers of science have about science, it can become a point of conflict and lead scientists to criticise philosophers for having an oversimplified concept of science.

2.3 A few words about objectivity

Since *scientific objectivity* was first established within science in the middle of the 19th century as a goal that scientists should strive towards (Daston and Galison, 2007, p. 27), it has become a rarely questioned concept in science. These days, the term *objectivity* is almost used like a 'magic word' in science. It is a word invoked by scientists, science communicators but also philosophers whenever they want to stress that science is something 'special', something that sets science apart from pure opinions. The *objectivity* of either research, researchers or lack thereof is a claim that is quickly resorted to in public debates about science. When something, a research result or a person or a method, is declared as *objective* in science, it is supposed to be a sort of seal of approval as Reiss and Sprenger note:

> Using the term "objective" to describe something often carries a special rhetorical force with it. The admiration of science among the general public and the authority science enjoys in public life stems to a large extent from the view that science is objective or at least more objective than other modes of inquiry. (Reiss and Sprenger, 2017)

In the context of public debates about science, the apparent *objectivity* of science is alternately used to 'prop up' scientific research results as irrefutable and significant or to undermine the work of the scientists by proclaiming that they are not 'objective'. At the same time, scientists themselves frequently 'conjure up' the term when they describe their own work or methods.

While the term *objectivity* is used often as the 'ultimate' signifier of 'good science', an in-depth examination of the use of the word *objectivity* in science quickly reveals that this is far from as clear-cut as the confidence, with which the term is applied. In fact, as Heather Douglas argues, *scientific objectivity* is "among the most used yet ill-defined terms in the philosophy of science and epistemology" (Douglas, 2004, p. 453).[11] In fact, a quick look at the literature also shows that there is not even a clear consensus among philosophers of science about the exact amount of definitions.

11 This also applies to the word *objectivity* in its historical evolution. Even though objectivity as an objective of science only has existed since approx. 1860, the history of the term *objectivity* itself goes further back and has quite a "somersault history" (Daston and Galison, 2007, p. 27), having changed its meaning almost to the contrary since first appearing in European languages.

Reiss and Sprenger, for instance, state that, in principle, there are two distinct basic concepts of scientific objectivity: *product objectivity* and *process objectivity*:

> According to the first understanding, science is objective in that, or to the extent that, its products — theories, laws, experimental results and observations —constitute accurate representations of the external world. The products of science are not tainted by human desires, goals, capabilities or experience. According to the second understanding, science is objective in that, or to the extent that, the processes and methods that characterize it neither depend on contingent social and ethical values, nor on the individual bias of a scientist. (Reiss and Sprenger, 2017)

The authors emphasise that particularly *process objectivity* comes in a variety of forms depending on what kind of scientific *process* (e.g., the structural organisation of science and methods of measuring) is meant to evoke objectivity.

In a similar, but somewhat more specific, fashion, Martin Carrier (2013, 2010) also distinguishes two different and contrasting concepts of scientific objectivity. He argues that concepts "of scientific objectivity are governed by two ideal types, namely, objectivity as adequacy to the facts and objectivity as reciprocal control" (Carrier, 2010, p. 207). He traces the former back to Francis Bacon, according to whom the ideal scientists is detached and neutral (Bacon, [1620] 1863; Carrier, 2013, p. 2549, 2010, pp. 207–208). The second meaning of objectivity is pluralistic in its approach. From this point of view, objectivity is reached through a diversity of points of view. This understanding of scientific objective has been popularised by Helen Longino (1990) but Carrier also sees familiar elements of this approach in the works of Karl Popper and Irme Lakatos (for a more detailed discussion of the pluralistic approach to objectivity and values, see Chapter 3.1.1).

Contrary to these dualistic definitions of *scientific objectivity*, other philosophers have argued that a more differentiated categorisation is more appropriate. While Megill (1994) argues that there are "four senses of objectivity" in general: *absolute, disciplinary, dialectical* and *procedural*, Heather Douglas (2004) finds at least eight different use cases of the phrase *scientific objectivity* divided into three different modes: the first type of objectivity is defined by the interaction of the scientists with the world. Secondly, there is an understanding of objectivity that is characterised through the personal (the value-related) reasoning process of the scientist. The third type of objectivity is based on the social structures and procedures of science.

The different and contrasting interpretations and categorisations of the phrase *scientific objectivity* indicate that the term is "not logically reducible to one core meaning" that simply (Douglas, 2004, p. 455).

It, therefore, seems prudent to follow Douglas here, who argues that these different types of objectivity are neither necessarily reducible nor incompatible. On the contrary, there is often more than one meaning at play when we call something *objective*.[12] Furthermore, the way we use the phrase *scientific objectivity* might change in the future. Certain definitions might be found wanting and others might newly emerge. We are by no means "finished developing the term" (Douglas, 2004, p. 468).

This diversity of meanings should also be kept in mind when we come across the term *scientific objectivity* in the next chapter. The simultaneously wide but not very well defined application of the term is reflected in the context of public climate-science debates. It is, therefore, not surprising that the term *objectivity* will also return in the following discussion about traditional ideals about science which are in conflict with how modern science actually functions. The most striking occurrence of this is in the context of the value-free ideal because in the public discourse objectivity is often used synonymous with the value-freeness of science (see Chapter 3.1). In the next chapter it will become clear why science cannot be *objective* in this sense.[13] Further, the phrase *objectivity* will reappear in connection with the ideal that observations in science can provide irrefutable evidence for or against a theory. In this context observations are commonly treated as *objective* in the sense that they are considered to be not open to interpretation (see Chapter 3.2). Contrary to this, many philosophers of science agree that observations have to be treated as theory-laden and theories are underdetermined by data. Particularly in climate science where a large amount of data has to be dealt with, models are usually considered to be "data-laden" and observations "model-filtered" (Edwards, 1999).

---

12 This, Douglas argues, also applies to the term *subjectivity* which has a considerable and irreducible variety of interpretations. Nor can subjectivity be seen as just the opposite to objectivity: "subjectivity is not just the lack of objectivity, and objectivity is not just the overcoming of subjectivity. Both are rich concepts, elements of which may be placed in stark opposition to each other" (Douglas, 2004, p. 470).

13 For a more successful application of a concept of scientific objectivity (following Longino, 1990, 2002) to aspects of climate science, see Leuschner (2012a) and Chapter 3.1.3.4.

At the same time the word *objectivity* is also used by climate scientists themselves. Here the term usually has a much narrower application and is commonly located in recurring debates about whether so-called *subjective* 'manual' procedures can be substituted by *objective* mathematical procedures. Here, *subjective* has a somewhat negative connotation. There are, for instance, discussions about whether the process of tuning can be made more objective by implementing an automated process of "find[ing] optimal sets of parameters with respect to certain targets" (Mauritsen et al., 2012, p. 16). In a similar vein, certain methods of data-processing (see Chapter 3.2) and procedures of model intercomparison (see Chapter 3.3), applied to study uncertainties in the models, are sometimes described as *objective* because there is an automated, mathematical element to the method. In that sense, the use of the term *subjective* in describing certain features of scientific methods requiring the specific skill and experience that scientists develop through their work will also be important to understanding the rising relevance of this experience to epistemological questions, as will be discussed in Chapter 3.4.

3. Three ideals of science

In this chapter three ideals about scientific methods, procedures, objectives and what constitutes good science, which are assumed to be widespread in the public understanding of science, are explored in relation to climate science. However, I would like to make clear here that the aim is not to present some kind of sociological study that makes some general assessment of the public understanding of science and from there establishes and categorises these ideals. Instead, I will take a top-down approach to this question and will infer from a number of instances from the history of climate science – as established in the introduction to this book – where science sceptics were very effective in discrediting particular climate research in the eyes of the public, that they benefited from the general popularity of certain idealised assumptions about how science operates. The primary objective of this chapter is to show, based on the work of philosophers of science over the last century, why a failure of science in general and climate science more specifically to live up to these ideals is not a sign of inadequate science. Further, it will be analysed why the failure of these ideals becomes particularly visible in the context of climate science.

3.1 Value-free science

3.1.1 Introduction: values in science

Science as a value-free endeavour has long been and continues to be an ideal upheld by the public, scientists and (to a certain degree) philosophers alike as a definition of what constitutes good science. This is no different when it comes to climate science. As we have already seen in the introduction to this book, the accusation that climate scientists are biased and, thus, not objective in their research is at the core of many climate-change deniers' arguments.

The value-free ideal of science is often intertwined with a certain understanding of *scientific objectivity*, where objectivity is defined by the relationship between science and (social) values. This kind of objectivity may take several different forms. For a better understanding of how varied scientific objectivity can be interpreted in relation to values, it is worthwhile to take a closer look at this type of objectivity in Douglas's classification of scientific objectivity introduced in the last chapter (2004). Douglas argues that there are (at least) three versions of this kind of scientific objectivity. The first option to define objectivity in this way (that is, in relation to values) is to claim that values should not be allowed to override evidence. This, as Douglas calls it, *detached objectivity* is by comparison a rather broad definition. It does not completely rule out that there is an appropriate role for values in science. However, there is also a narrower, more common understanding of scientific objectivity which does exclude values in (almost[1]) any form. This *value-free* ideal is what I will primarily discuss in this chapter.

The third kind of value-related understanding of scientific objectivity which Douglas identifies is that of *value-neutral objectivity*, a view of science, which acknowledges values in science (to a certain degree), but scientists are urged to take a middle-ground position. This is a point of view on the role of values in science that will not be discussed here in more detail. Suffice it to say that taking no sides at all might be undesirable in certain situations, if what lies on one side of the value spectrum is otherwise considered absolutely unacceptable, such as racist or sexist positions.

In the following, I will discuss why the value-free ideal cannot be maintained in the case of climate science and science in general. To that end, I will first outline the historic background. A look back in history helps to better understand the value-free ideal in general and how it has risen to such prominence in the last century. Before actually turning to the debate about the role of value judgements in climate science, I will also take a closer look at the discussion of value judgements in the context of inductive-risk assessments, which has taken up a prominent place in philosophical debates about the role of values in climate science.

---

1 Philosophers of science who advocate for this strict value-free ideal of science commonly acknowledge that there is a small number of "epistemic" values. These are considered to have an appropriate role in science compared to so-called "non-epistemic" values which are generally, according to this view of science, considered inappropriate; a distinction which will be further discussed in this Chapter.

3.1.1.1 The rise and fall of the value-free ideal

Historians and philosophers retrace the origins of the separation of values and science to Francis Bacon and the beginning of modern science and philosophy (Carrier, 2013; Douglas, 2009; Proctor, 1991). Bacon voices concern that moral deliberations would deceive men in their pursuit of scientific knowledge and prevent them from fully dedicating themselves to the advancement of science (Bacon, [1620] 1863). Another forerunner of the value-free ideal, which is often cited and should at least be mentioned here, is David Hume's distinction of "ought" and "is" and the notion that "ought" cannot be inferred from "is" (Hume, [1739–1740] 1888, p. 469).[2] However, it was not until the late 19th century and the rise of social science as a scientific discipline that the notion of value-free or value-neutral science, as we know it now, emerged (Proctor, 1991, p. 65). At the beginning of the 20th century the value-free ideal was prominently supported by the German sociologist Max Weber. Weber was the leading intellectual in the *Werturteilsstreit* advocating for a strict separation of science (specifically social science) from values. He was especially concerned about the university professor who might push their political ideals onto their 'defenceless' students and pass them on as scientific facts. Value judgements can also have an undesirable effect on science itself, according to Weber, to the extent "that whenever the man of science introduces his personal value judgment, a full understanding of the facts *ceases*" (1946, p. 146).

But even though the value-free ideal had a prominent and committed advocate in Weber, it did not prevail until the middle of the last century. As Douglas (2009, pp. 44–46) has shown, there was still a lively debate about the relevance and necessity of values in science and what form a value-free science

---

2 Proctor (1991, p. 65) remarks that Hume was not the first to make this distinction. He also notes, referring to Hampshire (1949) and MacIntyre (1959), that Hume himself did not fully separate "ought" from "is" and only ever meant that the former could not be derived logically from the later. "But ideas live a life apart from the intent of their authors. In the mid-twentieth century, Hume's call for a separation between "ought" and "is" became a rallying cry for scientists and philosophers defending the neutrality of science" (Proctor, 1991, p. 61). Proctor also stresses that Bacon's and Hume's position here must be seen in the context of a wider move from philosophers and scientists to separate science from religion and questions of ethics as well as a new recognition of subjectivity in science, which is most visible in the distinction between primary and secondary qualities (Proctor, 1991, p. 54).

should or could even take throughout the first half of the 20th century. Prominent philosophers of that time such as Robert Merton and Ernst Nagel voiced at least mixed feelings about the topic.[3] Others even argued that values in science were unavoidable or even necessary.[4] Particularly strongly debated were value judgements in the context of inductive risks. Richard Rudner (1953) and C. West Churchman (1948) argued that under certain circumstances scientists cannot but must make value judgements in their scientific deliberations. They note that whether or not scientists accept or reject a hypothesis depends not just on the epistemic evidence but also on the severity of possible social and ethical consequences of a wrong decision (a longer discussion of the inductive-risk argument follows in Chapter 3.1.2). Popular counterarguments against this reasoning are that such value judgements should be handed over to the public, while the role of the scientists is only to ascribe probabilities to hypotheses (Jeffrey, 1956), or that all value judgements scientists have to make can be solely determined by inner-scientific "canons of interference", i.e., epistemic values that are the same for all members of the scientific community (Levi, 1960, p. 356).

Though there were still some debates about the proper role of values in science at the middle of the last century, the debate, at least in the USA, soon died out at the beginning of the 1960s in favour of the value-free ideal. While there were still some discussions ongoing about this topic elsewhere in the world,[5] it is worthwhile to consider why the value-free ideal spread so quickly in the USA

3 Despite arguing for disinterestedness as an "ethos of science", Merton also sees science as embedded in a wider societal context in such a way that scientists have to consider the social and ethical implications of their work (Merton, 1973, pp. 267–278). Douglas also points out that science "'being value-free' is nowhere among the norms" (2009, p. 46) put forward by Merton.
Nagel (1961) also discusses the impact of values on sciences in an inner- and outer-scientific context in a variety of ways, though he does not regard the influence of values in science as far reaching as Rudner and Churchman do. A similar point of view is voiced by Hempel (1965); see also Douglas (2009, pp. 58–59).

4 Other examples of prominent philosophers of the time who, Douglas notes, did not advocate for thinking about science as a fully value-free realm are, e.g., John Dewey and Rudolf Carnap (Douglas, 2009, p. 47).

5 One might consider for instance, the second Werturteilstreit in Germany, which went on until the 1970s, and focussed on the role of values in sociology.

in the 1960s. In the post–World War II period the centre of discourse in philosophy of science, as much as science itself, shifted to the United States, which had, therefore, a global sphere of influence that did not end at the boarder. Further, it shows impressively how external (value-laden) pressure can influence the direction science and philosophy of science takes.

Douglas identifies two factors why the value-free ideal gained so much in popularity so quickly in the United States in the 60s. First, during the 1950s, the political situation in the USA had shifted in a way that put increasing pressure on all academics to distance themselves from anything that could be construed as support for communist ideas. Marxist philosophy traditionally sees science as situated in society so that the social and the scientific are interconnected. Fuelled by the McCarthy-era paranoia, many philosophers of science gave any position that could be misunderstood as political a wide berth.

This, Douglas notes, went hand in hand with another shift in philosophy of science already discussed in more details in Chapter 2.2: the separation of *context of discovery* from *context of justification*. Following Reichenbach's reasoning (1938), the former was deemed to be philosophically uninteresting and to be a topic of discussion for sociologists and psychologists but not philosophers.[6] Philosophy of science instead is supposed to focus on the logical justification of the result of scientific research. This distinction demands a restriction of the scope of research for philosophers of science, which was commonly interpreted to exclude a discussion about (social) values in science. The political situation made it attractive for philosophers of science in the USA to abandon any wide reaching, non-specific discussion about science in wider social context, Douglas argues, and instead "to professionalize their field, narrowing their expertise and focusing on a well-defined topic" (Douglas, 2009, p. 49).

The second factor Douglas cites as reason for the advancement of the value-free-ideal is the influence the publication of Thomas S. Kuhn's *The structure of scientific revolution* in 1962 had on philosophy of science. While the book had been hugely influential on philosophy of science in general, it also influenced the

6 In this context it might be interesting to note that Don Howard (2006) argues that Reichenbach's DJ distinction has to be interpreted as a way for Reichenbach to directly distance himself from Otto Neurath and the idea that values have a legitimate role in science. According to Howard, Reichenbach's position is an attempt to exclude the question of values by reducing philosophy of science to pure logic, whereas Neurath saw certain values as an unavoidable byproduct of the underdetermination of theories in science.

debate about values in science, according to Douglas. In the book Kuhn places science in its own historical context but also situates science outside of society. The separation from society is what makes (natural)[7] science, for Kuhn, such a successful endeavour:

> the insulation of the scientific community from society permits the individual scientists to concentrate his attention upon problems that he has good reason to believe he will be able to solve. Unlike the engineer, and many doctors, and most theologians, the scientists need not choose problems because they urgently need solutions and without regards for the tools available to solve them. In this respect, also the contrast between natural scientists and many social scientists proves instructive. The latter often tend, as the former almost never do, to defend their choices of a research problem – e.g., the effects of racial discrimination or the causes of the business cycle – chiefly in terms of social importance of achieving a solution. Which group would one then expect to solve problems at a more rapid rate? (Kuhn, 1962, p. 163)

By the mid-1960s the value-free ideal had truly become mainstream. Douglas writes that even in post-McCarthyism times, the ideal of value-free science was very attractive to science and philosophy of science alike for several reasons. For instance, the notion of the intrusion of values into science might 'bring back bad memories' of periods in the history of science where unwarranted outside forces interfered (see also Rudner, 1953, p. 6), or there might be concern that "science will lose its general public authority if a role for social or ethical values is admitted" (Douglas, 2009, p. 79).

The discussion about the role of values in science only gained momentum with the emergence of feminist philosophy of science. Feminist philosophers (for

---

7 Kuhn distinguishes here between natural and social sciences and sees the latter much more situated in a social context than the former. This, according to Kuhn, already transpires in the way that the training for future scientists is structured in the different disciplines: while the social scientists are required to study the original sources of previous scholarly disputes, where they learn to see different perspectives and arguments, the student of natural sciences is presented with condensed versions of research results from textbooks. Only advanced students actually study research papers directly. This "rigid education", according to Kuhn, prepares the young scientists optimally (in a period of normal science) for a (professional) life of puzzle solving (Kuhn, 1962, pp. 164–165).

example, Helene Longino, Donna Haraway and Sandra Harding) raised concern that there is no such thing as an objective, value-free "gaze from nowhere" (Haraway, 1989, p. 581).[8] Rather, they argue, upholding the value-free ideal would only hide the actual value judgements that come into play when doing science. Instead, feminist philosophers have promoted an honest handling and open communication of (potential sources of) value-influence in science. A popular proposal made by feminist philosophers of science to counterbalance inadvertent value-laden background assumptions is a more pluralistically and diverse organised science community (Longino, 1990). Although the feminist criticism of the value-free ideal was first met with reservation by conventional philosophy of science (see for example Kitcher, 1993), in the last two decades, the thinking that value judgements are inevitable in science has come back into the mainstream of philosophy of science. Particular the rediscovery of Rudner's 1953 paper has reactivated research interests into the question if and when values are an appropriate feature of science (e.g., Carrier, 2013; Douglas, 2009; Wilholt, 2009).

3.1.1.2 Epistemic versus non-epistemic values

With the rise of the value-free ideal another distinction also arose: that between *epistemic* and *non-epistemic* values. While many philosophers argued for value-free science, they also recognised that scientific theories on their own are underdetermined (see Chapter 3.2.2). To choose between two competing theories, more than empirical evidence is required. The same empirical evidence can support several even contradicting theories. Scientists, thus, need something more to make decisions between two or more equally well-established theories; some kind of value judgement is needed. Some philosophers have, therefore, suggested to make a distinction between appropriate, inner-scientific values and those values that come from an outer-scientific realm that the advocates of the value-free ideal try to keep out of science. Going back to Ernan McMullin (1982), these two types of values are often referred to as *epistemic* and *non-epistemic* values. More specifically, McMullin refers to five criteria defined by Kuhn, which are to fulfil the role of epistemic values in science: accu-

8 Ironically, many of those who first advocated for a value-free science at the turn of the last century made the (arguably value-laden) judgement that objectivity and value-neutrality were male attributes and women were, therefore, deemed not suitable for science (Proctor, 1991, p. 119).

racy, consistency, scope, simplicity and fruitfulness (Kuhn, 1977, p. 322).[9] Proponents of this distinction argue that epistemic values are distinct from non-epistemic ones because they "are presumed to promote the truth-like character of science" (McMullin, 1982, p. 702). Though they are "normative principles [...] given an initial commitment to these principles, the scientist need not and should not let his values, attitudes, and temperament influence his inferences any further" (Levi, 1960, p. 346).

Thus, the value-free ideal as it manifested itself in the early part of the second half of the last century is not a complete rejection of any kind of values, as Douglas notes:

> The ideal that has held sway since 1960 is a complex one. It does not hold that science is a completely value-free enterprise, acknowledging that social and ethical values help to direct the particular projects scientists undertake, and that scientists as humans cannot completely eliminate other value judgements. However, the value judgments internal to science, involving the evaluation and acceptance of scientific results at the heart of the research process, are to be as free as humanly possible of all social and ethical values. Those scientific judgements are to be driven by values wholly internal to the scientific community. Thus the value-free ideal is more accurately the "Internal scientific values only when performing scientific reasoning" ideal. (Douglas, 2009, p. 45)

By introducing this division between 'good' epistemic values and 'bad' non-epistemic, social values, the image of science as a space free from at least external scientific influences appears to stay intact, even if one recognises the issue of underdetermination.

But in what sense are these epistemic values actually distinct from non-epistemic social, ethical or political values? Critics of this dichotomy have argued that many of the values that are traditionally seen as purely epistemic do not strictly say anything about the truth of a theory (Laudan, 2004). For instance, a theory can be true even when there is an alternative one with a wider

9 Although Kuhn states that these five criteria are not exhaustive, they are often treated as the canon of epistemic values by philosophers of science. Rooney shows how philosophers make slight differences in how they define and name those criteria but finds that the "fact that there is no clear consensus about what is included among the epistemic or constitutive values does not overly concern many of those who make the distinction" (1992, p. 14).

scope. Similar things can be said about the values of *simplicity* or *fruitfulness*. That does not mean that these kinds of values are of no worth to science. There is more to a 'good' theory than accuracy, as Laudan with reference to Bas van Fraassen points out:

> Bas van Fraassen famously argued that a theory does not have to be true to be good. We can add to that dictum a new twist: a theory does not have to be false to be bad. A theory may be bad because it fails the test of possessing the relevant nonepistemic virtues. In other words, we expect our theories to do much work for us, work of a sort that most merely true statements fail to do. (Laudan, 2004, p. 19)

Nor are epistemic values as universal as their advocates claim. Helen Longino (1990, pp. 83–102, 2002, 2008) argues that one could easily imagine a set of alternative epistemic values, which would also fill the gap left by the underdetermination of theories. One might, for instance, substitute the values of *simplicity* and *scope* by *heterogeneity* and *mutuality*. There is, after all, "no prior reason to think the universe simple, that is, composed of very few things" (Longino, 2008, p. 73).

Besides *empirical adequacy*, Longino proposes feminist philosophers might advocate for alternative epistemic values such as *novelty*, *heterogeneity*, *mutuality* and *decentralisation of power* because they support feminist objectives in science. Longino provides several examples from the history of science to argue which specific values are observed can have actual social consequences. Such instances are medical research only done on white males or economic theories of the household assuming patriarchal structures, both adhering to an ideal of simplicity (Longino, 2008, pp. 74–75).

Therefore, argues Longino, these kinds of values – the traditional epistemic values and any kind of alternative values – are actually heuristics. These heuristics are specific to particular scientific communities and might shift over time when they no longer serve their purpose. Thus, the alternative values Longino proposes are not feminist in themselves, even though feminist might favour them but "subordinat[e] to a broader cognitive goal" (2008, p. 78). Ergo, there is nothing special about so-called 'epistemic' values:

> The feminist and traditional virtues are on par, epistemologically. Both have heuristic but not probative power. As heuristics, they help an investigator identify pattern or order in the empirical world. They are often transmit-

> ted as part of an investigator's training, as part of the common, taken-for-granted-background. (Longino, 2008, p. 74)

Hence, opponents to the value-free ideal have argued that the alleged epistemic values are not purely epistemic in nature but emulate current dominant social values. Feminist philosophers, for instance, have demonstrated in case studies from the fields of biology, primate studies, economics and medicine (amongst others) how androcentric values are sometimes 'hidden' behind the apparent 'objective' epistemic values.[10] Because of this some philosopher such as Laudan (2004) and Douglas (2009) have argued that the term *epistemic* should be dropped completely to describe values that help scientists in their reasoning process and that the more appropriate word for these kinds of values would be *cognitive*. *Cognitive* then refers to "those aspects of scientific work that help one think through the evidential and inferential aspects of one's theories and data. Taking the label *cognitive* seriously, cognitive values embody the goal of assisting scientists with their cognition on science" (Douglas, 2009, p. 93). Thereby, cognitive values function as an "insurance policy" (Douglas, 2009, p. 107); they increase the likelihood to find possible mistakes in the reasoning process.

But what about those virtues that are actually truth-conductive? Laudan argues that 'true' epistemic values are a small subsection of the cognitive values (2004, p. 19). Douglas, however, suggests actual epistemic virtues such as *empirical adequacy* or *internal consistency* should be viewed not as values but as necessary criteria for any kind of scientific theory (2009, p. 94). For example, it seems difficult to imagine a satisfying scientific theory that is not internally consistent.

This way of defining the role of values in science paints a different picture than the value-free-ideal. There is no longer a clear separation between 'acceptable' epistemic values and 'not acceptable' non-epistemic values. Douglas argues one should instead imagine the different types of values as different areas on a landscape that might intersect in certain places (Douglas, 2009, p. 91).

In my opinion Longino, Douglas, Laudan and others have raised valid concerns in respect to the dichotomy of *epistemic* and *non-epistemic* values. But one might as well question whether the term *cognitive* is not equally badly defined and far-reaching. Moreover, one might even ask whether it makes much sense to make

10 See, e.g., Haraway (1989); Keller (1985); Longino (1990). For an overview, see also Douglas (2009); Longino (2008); Rooney (1992).

this distinction at all if one takes Longino's claim serious that cognitive values might actually be based on concepts external to science. I will, however, not discuss this any further here as that controversy would lead us astray.

In order to avoid confusion over what the terms *epistemic* and, therefore, *non-epistemic* values actually refer to, I will in the following use the term *social and ethical values*. The allegations that climate scientists are biased and the subsequent philosophical debate about the role of values in climate modelling focuses exactly on those values; the kinds of values that proponents of the value-free ideal fear would 'contaminate' scientific research. The question what actually the appropriate place for values in climate science is and how severe the value-ladenness of climate science actually is centres on *social* and *ethical* values. The role of *epistemic* or *cognitive* values in climate science, on the other hand, is normally not disputed. Though there are differences between *social* and *ethical* values (Douglas, 2009, pp. 92–93) in general, I will not discuss these any further here. As this chapter will show, the practice of climate modelling involves a myriad of epistemically not fully constrained decisions and it is often impossible to determine retroactively if and what values were relevant in making those decisions. Thus, the question will be to what extent it can be argued that these kinds of possible value-laden judgements are not inappropriate (as the value-free ideal asserts) rather than what constitutes the specific value. For this reason I will refer to *social values* in short for *social and ethical values* as the kind of values that are contrary to cognitive values and according to the value-free ideal do not have a place in scientific reasoning processes.

Much of the current discussion about the relevance and unavoidability of social values in science, including climate science, is built on what is known today as the *argument from inductive risks*, specifically two forms of the argument made by Rudner (1953) and Douglas (2009). Which is why I will now discuss the inductive risk argument in more detail before turning to the specific issues raised by philosophers of climate science. In the following discussion of Rudner's and Douglas' remarks on value judgements in the context of inductive risks, I will specifically use the term *ethical* value where it supports the argument – contrary to the later discussion of the situation in climate modelling, where such a differentiation is, as said above, not particularly helpful, (otherwise I will continue to use the short version of *social values*, where both types of values are concerned).

3.1.2 Inductive risks and social values

Carl Hempel, who first introduced the term, defines inductive risks as the risk of accepting an empirical law to the extent that "the presumptive law may not hold in full generality, and that future evidence may lead scientists to modify or abandon it" (Hempel, 1965, p. 92). In these situations, opponents of the value-free ideal contend that scientists sometimes cannot but must make social-value judgements.

In his widely discussed 1953 paper "the Scientists qua Scientists makes value judgments" Richard Rudner argues that contrary to what the value-free ideal proposes – that there is an appropriate, even necessary role for social-value judgements within science, under the assumption that accepting or rejecting hypotheses is a quintessential part of the scientist's work.[11] In a nutshell the argument goes like this: when one accepts this premise and also assumes that ultimately a scientific hypothesis can never be fully verified, then one also has to assume that "in accepting a hypothesis the scientist must make the decision that the evidence is *sufficiently* strong or that the probability is *sufficiently* high to warrant the acceptance of the hypothesis" (Rudner, 1953, p. 2). In those cases where the research objective has a wider social application, judging whether or not a hypothesis is sufficiently proven has to be done under consideration of social values. Rudner demonstrates this with the now often quoted example of the testing of drugs versus belt buckles before they are released for sale. We require a much higher standard before we accept the hypothesis of a drug containing a toxic ingredient, only in such a quantity that it is still safe to use than we require of a load of belt buckles based on a sample size because the stakes in the first case are so much higher. Put slightly differently, compared to the consequences of falsely accepting the hypothesis that a drug is safe to use, a not properly working belt buckle might be annoying or embarrassing but has no (or only under exceptional circumstances) possible deadly consequences. Thus, a much more thorough testing is required when it comes to medicine than clothing accessories. This difference in requirements for testing standards arises not because of epistemical but ethical considerations. For this reason, Rudner concludes that the scientist cannot avoid value judgement in their role as a scientist.

11 Similar opinions have been voiced by others at the time, such as Churchman (1948) and Frank (1953). However, Rudner's paper has been the one discussed the most in subsequent years.

One might now be inclined to argue that the scientist can nevertheless hand over the responsibility of making value judgements to the public. This position, commonly ascribed to Richard Jeffrey, who claims that the scientist themself does not have to accept or reject a hypothesis but merely has to disclose the probability for one. According to Jeffrey, the scientist, after all, could not possibly have insight into all the possible consequences of their acceptance or rejection of a hypothesis:

> One cannot, by accepting or rejecting the hypothesis about the polio vaccine, do justice both to the problem of the physician who is trying to decide whether to inoculate a child, and the veterinarian who has a similar problem about a monkey. To accept or reject that hypotheses once for all is to introduce an unnecessary conflict between the interests of the physician and the veterinarian. (Jeffrey, 1956, p. 245)

Thus, following Jeffrey, the scientists are in no way forced to make judgements about the acceptability of a hypothesis. This is something that can be passed on to those who apply the information they get from the scientists in practice.

Rudner, on the other hand, argues that even just stating a probability for a hypothesis requires value judgements, as stating a probability *p* for the occurrence of a hypothesis *H* requires "*the acceptance by the scientist of the hypothesis that the degree of confidence is p*" (1953, p. 4). That is, coming to the conclusion that the probability of *H* is *p* requires accepting a further hypothesis *H'*. Therefore, a common counterargument to Jeffrey's position is that the problem of the value judgements is just transferred to another level.

The disagreement between Rudner's and Jeffrey's position regarding the need for scientists to make value judgements is often framed as a disagreement over the question of whether or not probabilities attributed to hypotheses are actually something that can be accepted or if they constitute a degree of belief in a conventional Bayesian sense (Steel, 2015; Winsberg, 2012). According to the latter position, personal probabilities are not something to be accepted but something one has, which may shift in light of new evidence but does not usually involve a conscious decision process. From this point of view the counterargument to Jeffrey that noting probabilities requires acceptance of those does not hold up. However, it seems questionable to what extent scientists can actually be described as perfect Bayesian actors. Steel (2015), for instance, points out that scientists often hold vague degrees of belief. In Chapter 3.1.3 we will return to this issue with a climate-science specific argument first made by Winsberg (2018, 2012).

As discussed in Chapter 3.1.1.1, the position of (social-)value judgement having no place in the scientist's everyday life as scientists, prevailed in philosophy of science in the subsequent years. In the last few decades, however, Rudner's paper emphasising the importance and unavoidability of social-value judgements in cases of inductive risk assessment in science has seen a resurging interest. Specifically noteworthy here is Heather Douglas's (2000) account of how scientists often are confronted with a number of methodological decisions that require social value judgements even before they accept or reject a final hypothesis. While Douglas does not, as has been pointed out, directly refute the Bayesian interpretation of Jeffrey's argument (Steel, 2015; Winsberg, 2012, 2018, pp. 135–136; Parker and Winsberg, 2018), it is still worthwhile to take a look a Douglas' reasoning for two reasons. First, if one takes scientists not to be perfect Bayesian actors, then it indicates that value judgements are a common and necessary element to decision-making processes in science way before scientists have to decide whether or not to accept the hypothesis. Second, Douglas also concludes from this that one can assess the appropriate place of value judgements in science by distinguishing between a direct and indirect role of values in science, a question we will also return to.

3.1.2.1 Social values and methodological considerations

Scientists do not only deal with inductive risks and thereby value judgements when evaluating hypotheses at the end of a scientific project. Douglas identifies three stages at which, she argues, social values can and do have a legitimate role in science-internal processes due to considerations of inductive risks[12] – which is not limited to the final evaluation of the hypothesis:

> If one follows the general schema of the methodology from a scientific research paper, significant inductive risk is present at each of the three "internal" stages of science: choice of methodology, gathering and characterization of the data, and interpretation of the data. At each point, one can make a wrong (i.e., epistemically incorrect) choice, with consequences following

12 Note that Douglas' definition of the term *inductive risks* here is rather board and goes beyond the risks of accepting or rejecting a hypotheses, but refers to errors more generally, i.e., making a "epistemically incorrect" decision at "'internal' stages of science" (Douglas 2000, p. 565). Harvard and Winsberg (2022) criticise this broad definition of inductive risks and argue that differentiating between *inductive risks* and *representational risks* is useful, particular in the context of scientific modelling (see Chapter 3.1.3.1).

from that choice. A chosen methodology assumed to be reliable may not be. A piece of data accepted as sound may be the product of error. An interpretation may rely on a selected background assumption that is erroneous. Thus, just as there is inductive risk for accepting theories, there is inductive risk for accepting methodologies, data, and interpretations. (Douglas, 2000, pp. 564–565)

Douglas rejects the assumption that in these cases the task of making value judgements can simply be handed over to the public. She illustrates this with the example of a case study of animal testing and the risk of dioxin inducing cancer in rats.

First of all, Douglas argues that in the specific case study one methodological decisions scientists have to make concerns the level for statistical significance (2000, pp. 565–569). As the control group of rats will also show a natural amount of cancer, the scientist must decide on a standard for statistical significance, beyond which the amount of cancer found in a rat population is considered the result of the exposure to dioxin. If they choose not to go along with conventions of that particular research field,[13] Douglas notes, scientist then have to consider the consequences of false positive (that is falsely accepting a hypothesis) or false negative (erroneously rejecting a hypothesis) errors when defining the level for statistical significance. Depending on how low or high the standard for statistical significance is set, one risks more false positive or false negative results. When the toxicity is wrongly overestimated, it may have a negative impact on the affected industries, whereas when scientists underestimate the possible toxicity, serious consequences for public health may arise. Therefore, the decision to go one way or the other is a question of trade-offs. Douglas emphasises that one can only reduce both the risk of false negative and false positive errors under a significant increase of costs, that is, in this case improving the experiment by increasing the number of research objects. Thus, it comes down to the question how the scientists value the possible consequences of both options:

In finding the appropriate balance between false positive and false negative errors, we must decide what the appropriate balance is in the consequences

13 It seems nevertheless reasonable to note that even when scientists follow these conventions, one might as well argue that in those cases scientists do make a value judgement to go along with such conventions.

> of those errors: overregulation and underregulation. Selecting an appropriate balance will depend on how we value the effects of those two consequences [...]. Finding the balance requires, among other things, weighing the non-epistemic valuations of the potential consequences. (Douglas, 2000, p. 568)

That is, in situations of methodological uncertainty where the decision will have political and social implications scientists cannot but refer to social values, Douglas argues.

Secondly, when it comes to the characterisation of data, evidence is rarely unambiguous. In these instances and when the results have consequences for public safety, scientists likewise have to consider the risks of false positives and false negatives, Douglas notes (2000, pp. 569–572). In the dioxin cancer studies Douglas examines, there was a significant number of cases where the scientists did not agree if the rat liver slides showed cancerous lesions. Different groups of pathologists who had evaluated the same samples at different points in time came to different conclusions how those borderline cases should be classified. In some situations, as Douglas points out, these types of discrepancies can be circumvented to a certain degree by letting the pathologist examine the tissue sample 'blind', so that they do not know whether the samples come from the rats exposed to dioxin or the control group. Therefore, when scientists have an evenly distributed tendency to false positives (or negatives), the errors in judgement should (in theory) balance each other out. But this approach would not work for most borderline tissue samples in Douglas' case study as the liver tissue samples of rats having been exposed to a high level of dioxin would also show signs of acute liver toxicity, which the experienced pathologists evaluating the samples would recognize. Thus, Douglas concludes, inductive risks and associated social-value considerations also do play a role in data assessment in this particular case and beyond:

> This case demonstrates that there is inductive risk in how one applies categories used in data characterization and that such inductive risk can be linked to non-epistemic consequences. [...] The consequences of the errors are identifiable and need to be weighed in order to determine which errors are more acceptable. In other cases, inductive risk may be present in the selection of the categories to be used as well as the application of the categories in the characterization of the data. In addition, judgments are made in science concerning whether to keep data or whether to discard the data as unreliable. At all these decision points, there is the risk of error, and

> with that risk, the need to consider both the epistemic and non-epistemic consequences of error. (Douglas, 2000, p. 572)

The third instance, where scientists have to consider inductive risks in their daily work, is the interpretation of data, according to Douglas (2000, pp. 573–577). In the case of the dioxin-study, there was considerable disagreement whether the result should be interpreted as there being a specific threshold for the dose of dioxin after which it causes cancer or whether the response increases consistently with the dosage. Thus, Douglas argues scientists have to take the inductive risks into account and weigh the consequences of potential errors in their judgement. Depending on whether one chooses a *threshold model* or a *linear extrapolation model*, the acceptable dosage for human consumption will be set differently. The consequences of a linear extrapolation model are usually stricter regulations than of a threshold model. Which one is chosen has consequences for the general public and the industry more specifically, as a wrongly chosen threshold model will most likely have negative consequences for public health whereas if it turns out that one erroneously adopts a linear extrapolation model industries will most likely be overly regulated.

Thus, in the case of socially sensitive research, scientists have to make social-value judgements at different stages in the science internal process, concludes Douglas. Further, in those cases, where inductive risks are at play "value-free science is inadequate science" (Douglas, 2000, p. 559). And contrary to Jeffrey's claim, scientists are the most competent and often only option for making these decisions, Douglas claims:

> The most important reason is that it is doubtful anyone could fully take over this function for scientists. Because science's primary goal is to develop knowledge, scientists invariably find themselves in uncharted territory. While the science is being done, presumably only the scientist can fully appreciate the potential implications of the work, and, equally important, the potential errors and uncertainties in the work. And it is precisely these potential sources of error, and the consequences that could result from them, that someone must think about. The scientists are usually the most qualified to do so. (Douglas, 2009, pp. 73–74)

In an attempt to clarify the appropriate role for social values in science, Douglas also differentiates between the direct and indirect role of values (2009, pp. 95–108). At the early stage of a research project, value judgements often have an (acceptable) direct role, Douglas points out. There are, e.g., legitimate reasons

for ruling out certain methodological approaches for ethical reasons from the beginning. We do not, for instance, endorse certain kinds of experimentations on human beings for ethical reasons. In a similar fashion, ethical or social values may be relevant when determining the objective of a research project. For example, governments might be more inclined to fund those projects which have distinct social relevance. Douglas argues, in these sort of situations, at the beginning of a scientific project, social values may take a direct role in such a way that they "determine our decisions in and of themselves, acting as stand-alone reasons to motivate our choices" (2009, p. 96). Values having a place in establishing research objectives and setting ethical boundaries when it comes to methodology, are widely accepted – even by proponents of the value-free ideal – because these decisions are seen as still taking place at a stage of the research project that has a 'pre-scientific' character.

Much more contested is what Douglas identifies as the indirect role of values. It refers to the role of values in science we have primarily discussed so far. Douglas applies this term to those instances during an ongoing research project when scientists have to make decisions under uncertainty. They help review whether there is sufficient evidence considering the specific circumstances, make decisions and weigh the consequences of potential errors in judgement in the way discussed above. In this form, the role of values is contingent upon the specific evidence at hand, Douglas contends.[14] When new evidence reduces the uncertainty, it also reduces the need and the place for values.

Thus, Douglas argues, depending on what stage a scientific research project is at, values may take up different roles. At an early stage, values can legitimately direct our choices by putting value on it in itself, whereas during the research project values should only take an indirect role and aid scientists when they are facing uncertainties.

However, these roles are not as clear-cut as they might seem at first glance, Douglas concedes. Under certain circumstances, direct value judgements might be inappropriate, even at an early stage of a scientific project. These, notes Douglas, are those cases where "a direct role [...] undermines the value of science itself" (2009, p. 101). This might be the case when objective and

14 That the role of values should be limited to an indirect role, once a research project is under way, also holds for cognitive values, according to Douglas (2009, pp. 107–108). This will not be discussed in great detail here because social, not cognitive values are at the centre of the argument concerning value judgements in climate science.

methodology predetermine the result of a research project. Furthermore, Douglas concludes that under certain (exceptional) circumstances, it may be appropriate for values to interfere directly in the science-internal process. Such might be the case, when scientists have to adjust their methodology, because it turns out that the methodology chosen at the early stage of the research project is in fact ethically not acceptable. We will return to the question to what extent it makes sense to distinguish between a direct and indirect role of values in science at the end of Chapter 3.1.

3.1.3 Social values in climate science

Once we turn our attention to scientific disciplines which have to deal with additional epistemic challenges coming from the high complexity of the systems under investigation, such as climate science, it becomes clear that the role of values within the scientific process reaches even further than inductive-risk assessments. In such cases, contrary to what the proponents of the value-free ideal envisioned, the significance of social values cannot be reduced to the role of setting goals at the beginning of the model-building process. Nor can the role of social-value judgements at internal stages of the scientific process be cut down to, as Douglas argues, decisions under uncertainties. Rather, as Winsberg has shown, the possibility of social value-judgements lies deep within the "nooks and crannies" (2012, p. 132) of climate science.

What differentiates the cases of social values in science discussed here so far from climate science are the specific epistemic challenges rooted in the high complexity of both system and models which entails a great deal of epistemically not fully constrained decision making. Because of this, as we will see, when it comes to the relevance of value judgements, "predictive preferences" (Winsberg, 2012) gain in significance and "representational risks" arise (Harvard and Winsberg, 2022). In the following I will discuss the consequences of this for the ways that social values might interfere in the model-building process. It will be shown that the complexity of the climate models does not only make it impossible to rule out that some science-internal decisions were made under considerations of social values but even to retroactively (fully) disclose them. In fact I will claim, this complexity is the reason why we do not have to fear that bias and wishful thinking could impact the models in an epistemically untoward way.

3.1.3.1 Unconstrained decision making, predictive preferences and cost restrictions

As argued before (see Chapter 2.1), there is no such thing as a fixed construction manual for a climate model. On the contrary, there is a plurality of models concerning scale, complexity and objective (for an overview see Parker, 2018). As has been pointed out, considering all idealisations and trade-offs, the quality of these kinds of models can only be assessed with respect to their purpose (Chen et al., 2021; Parker, 2009). But even with models possessing a similar set of goals, scale and complexity, there are numerous ways to construct a global climate model. On a macroscopic level, Alexander and Easterbrook (2015) have shown that there are different modelling traditions in Europe and America, which are both epistemically equally well justified (see Chapter 2.1). The emergence of these kinds of epistemically unforced methodological questions are not singular events in the process of 'assembling' these kinds of computer simulations. It is not an uncommon occurrence in process of model construction that there are several different options how to represent one and the same climate mechanism within the model depending on the particular objectives and underlying modelling 'philosophies'. In this context scientists are also often faced with the question to what extent and in what way to include specific processes. How these questions are answered depends on the purpose of the model but also on cost-benefit deliberations.[15] For instance, one might imagine a hypothetical situation where three different modelling groups have to decide how to implement a climate-relevant process into their model, which can be represented either resolved or parametrised. One modelling group, for example, might choose to represent a specific process in their new model by relying on a parametrisation that they are well acquainted with, have used and tested in a previous model because they are content with the performance of the model with said parametrisation. A different research group might decide that it is worthwhile to invest in increasing the resolution of their new model so that it is possible to integrate that specific process in their model directly in a resolved way. The scientists conclude that this will hopefully result in a physically more accurate representation of the process. A third modelling group, however, might make the decision to improve the existing parametrisation in

15 'Cost' has to be understood in the broadest sense here, see below.

their current model because they think that that process is more effectively represented in the model in form of a new parametrisation.[16]

Similarly, it is widely acknowledged among climate scientists that there is a variety of ways of how to tune a model or model components. Tuning is the fixing of certain, otherwise not very well constrained parameter so that the overall model result fits better with scientists' expectation based on observations and expertise. Although there are some conventions, as it is not possible to tune a model perfectly in respect to every variable, differences in priorities and in well-established modelling cultures at individual institutes also influence what approach to the tuning process is taken (Chen et al., 2021, pp. 217–218; Mauritsen et al., 2012). Even though scientists will have good reasons for choosing the tuning method and objectives that they do, they are also aware that their decision are not strictly epistemically constrained (Hourdin et al., 2017).[17]

In practice, how these decisions are made is often also dependent upon the modelling culture at different research institutes and the specific histories of the models. However, it should be emphasised here that these are decisions where the scientists have good (epistemic and methodological) reasons for deciding the way they do. But if they had chosen one of the other paths, because of

---

16 In practice, as Helen Guillemot (2017) has shown, modelling groups only rarely invest in improving existing parametrisations because the costs in terms of time and effort are estimated to be too high compared to the benefit of the outcome. Even though a new parametrisation in isolation might seem as an improvement, due to compensating effects and tuning, it will inevitably perform worse than the old one when first integrated into the model, which means additional work. See also Chapter 3.3.3.2 and Chapter 2.1.

17 Historically, tuning has not been an issue that has been discussed much within the climate science community. Hourdin et al. (2017) point out two possible reasons for this. On the one hand, tuning may be considered to be somewhat 'unscientific' and "more engineering than science, an act of tinkering that does not merit recording in the scientific literature" (2017, p. 590). On the other hand, there may be concerns that emphasising the necessity of tuning may give 'ammunition' to climate change sceptics. So that the climate science community may see the whole process of tuning "indeed as an unspeakable way to compensate for model errors" (Hourdin et al., 2017, p. 590). However, in the last decade there have been several attempts to bring the actual reasoning process behind different tuning strategies to the forefront (Schmidt et al., 2017). Most notably, in the widely discussed paper from Mauritsen et al. (2012) the authors explore as a case study what the effects are of different choices made in the tuning process of their model.

different objectives or modelling cultures, that choice would have been equally well justified.[18]

Considering the specific epistemic challenges of climate-model development, particularly the difficulties in accessing analytical understanding (see Chapter 2.1) and the long timeframe, these are not necessarily decisions that can be made fully at an early stage when a new model is initiated. As will be argued in the following, these are choices that have to made and (re-)assessed continually during the process of constructing and evaluating climate models.

In the context of climate science the role of value judgements are mostly discussed in the context of what Eric Winsberg has termed "predictive preferences" (2012, p. 131, 2018, p. 138). As there is no such thing as a perfect climate model representing every aspect of the global climate equally well, climate models echo predictive preferences climate scientists have for specific variables or processes that they consider to be more significant for their research questions (Tebaldi and Knutti, 2007, pp. 2045–2055). Trade-offs have to be made with respect to which aspects of the climate system are to be prioritised, not just in respect to where one decides to invest time and money but also purely on grounds of the intricate characteristics of the model building process. Considering that the model show path dependency, specific modelling choices will restrict what further options are available and impact how well other aspects of the climate can be represented in the model. The setting of priorities is also a necessary feature of climate model tuning (Mauritsen et al., 2012).[19] One cannot tune the perfect model; a model can only be tuned well with respect to certain aspects. There are again trade-offs to be made to the

18 This is not unique to climate science; it also affects other scientific disciplines using computer simulations of a similar complexity. A similar situation is, for instance, described by Ruphy (2016, pp. 100–101) in relation to the use of computer simulations in astrophysics.

19 Hourdin et al. describe what kind of different objectives these might be and what influences them: "different models may be optimized to perform better on a particular metric, related to specific goals, expertise, or cultural identity of a given modelling center. Groups more focused on the European climate may give more importance to the ocean heat transport in the North Atlantic, whereas others may be more concerned with tropical climate and convection. Some groups may put more weight on metrics that measure the skill to reproduce the present-day mean climatology or observed modes of variability, while others may privilege process-oriented metrics targeting processes that are believed to dominate the climate change response to anthropogenic forcing" (2017, p. 592). Note that possible tuning goals listed here do not just concern

extent that tuning models to improve the representation of specific features of the climate often goes hand in hand with a decline in the model performance in respect to other aspects of the climate (Hourdin et al., 2017, p. 596). These value judgements concerning what to prioritise when tuning a model also cannot be circumvented by relying on algorithmic (so-called *objective*)[20] methods to find optimal parameters for a specific target, because it still requires the scientists to set goals for the tuning process. As Hourdin et al. point out: "An objective algorithm merely identifies those parts of the procedure that require the subjective scientific expertise of the modeler" (2017, p. 594).

There are further ways in which social-value deliberations might influence methodological decision-making in climate science. Often these kinds of decisions are determined by *considerations of costs*. Costs have to be understood here in a broad sense, it does not only include financial deliberation but also, for instance, questions in respect to time and effort put into developing a model or specific part of the model.[21] These considerations are similar to predictive preferences but are different in the primary goal, though a finite number of resources will eventually also lead to predictive preferences.

Harvard and Winsberg (2022) argue that in the context of computer modelling one ought to distinguish between inductive risks and representational risks. Representational risks go beyond the risk of upholding a false fact. Harvard and Winsberg emphasise that representational decisions are not questions of right or wrong but whether or not an adequate choice for the intended purpose and considering all relevant epistemic agents is made. This is an important distinction that does not just pertain to the here discussed occurrence of unforced decision making in climate modelling. Harvard and Winsberg note that in the context of complex computer simulations in general it is a well-established insight that they often include elements that do not have a direct representation in the 'real' world. It is not even uncommon in complex computer simulations that a 'false' parameter might be the right choice as it adequately compensates for inaccuracies elsewhere in the model. One example Harvard and Winsberg point out from climate modelling are parameter values for cloud formation,

pure scientific research questions but may also serve distinctly socio-political purposes concerning dealing with anthropogenic climate change.

20 See Chapter 2.3 and Chapter 3.4.3.

21 In science time and effort of can, of course, again be translated into financial costs (see also Knorr-Cetina, 1981, pp. 40–41).

which are chosen to balance energy leaks in climate models at the top of the atmosphere (2022, p. 15).

One might also ask the question to what extent different modelling cultures influence how epistemically unforced decisions are made. Different modelling groups or institutes can develop different traditions concerning central questions in the modelling process (Skelton et al., 2017). Besides the differences between European and American model structures (Alexander and Easterbrook, 2015) already discussed here climate scientists have also stressed that there are different approaches to tuning at different institutes (Hourdin et al., 2017; Mauritsen et al., 2012). There can also be diverging attitudes towards basic ideas about the future of climate modelling, e.g., whether to invest in better parametrisation or in reducing the grid size so as to reduce the dependency of models on parametrisations (Guillemot, 2017). This is similar to different 'lab cultures' which are traditionally attributed to groups of scientists working in laboratory settings (Knorr-Cetina, 1999; Latour and Woolgar, 1979).

At first glance these modelling cultures seem to resemble *cognitive* values, insofar as they seem not to be affected by non-science related assumptions. However, just as cognitive values might as well be fundamentally grounded in political or ethical ideals (Longino, 2008), it does not seem too far-fetched to question if this might not also be the case for these modelling cultures. The belief that scientists should invest more time and energy into improving parametrisation, for example, instead of being hell-bent on reducing the grid size of models could also be influenced by the hope to thereby provide better, policy-relevant results considering the time constraints.

While philosophers of science have discussed the role of values in climate modelling extensively, one question that has so far has seen much less attention is whether the collection and creation of observational data (sets) are similarly affected by social value deliberations. But it is to be assumed that the situation here is not much different to that in climate modelling. Observational data are often understood by laypersons as some kind of 'objective' benchmark against which the quality of a theory or a model can be assessed. However, as will be discussed in more detail in Chapter 3.2.3, climate data, just like models, are impacted by a wide variety of uncertainties and inaccuracies. Therefore, a great deal of processing in terms of filtering and homogenising has to be done to create global data sets. This comes hand in hand with some degree of methodologically not fully constrained decision making (Parker, 2018). As, for example,

the case of satellite data shows (Chapter 3.2.3.1.2) there is some wiggle room in how to interpret this kind of data, the possibility of some inductive risk consideration taking place does not seem to be too far of. Further, Brönnimann and Wintzer (2018) point out the context-dependency of climate data. That is, "climate data products carry imprints of social, political and economic contexts" (Brönnigmann and Wintzer, 2018, p. 4) of the circumstances under which they were created. One place they note this can be observed is in the historical inequality in climate data coverage, which they argue "is not just a data problem, but also one that affects climate justice" (Brönnigmann and Wintzer, 2018, p. 4). As will be further explicated in Chapter 3.2, climate models are not fully theoretical constructs but are "data-laden" (Edwards, 1999), so what data is collected and available also has an impact on the model-building process. Thus, it has to be assumed that the decisions made in the process of the creation of data sets (which do, as can be derived from Brönnimann and Wintzer's analysis often resemble predictive preferences) are anchored deep within the "nooks and crannies" (Winsberg, 2012, p. 130) of climate modelling.

Now one might be inclined to further discuss in what particular ways social values are relevant in specific climate-science internal processes and what their specific impact is. But, in my opinion, that would be somewhat missing the point. What is of relevance here is not which specific values influence climate modelling, but that there are, as Winsberg has pointed out, literally "thousands of unforced methodological choices" (2012, p. 130), which require to set priorities that cannot be determined purely on the basis of epistemic considerations alone.

3.1.3.2 Non-traceability

In the debate about values in science, it has been argued that in those cases where scientist have to make judgements on the basis of social-value deliberations scientists should take care of communicating what went into these decisions as explicit as possible to policy makers and the general public to ensure the integrity of science (Douglas, 2009, p. 136). Against the backdrop of the vast number of epistemically unforced decisions and the different ways in which social values may play a role in the construction of a climate model, the question is whether this call for disclosure of all possible value-laden assumptions

can still be met.[22] What complicates this even further is the fact that complex, global climate models are not built by just one scientist or even one research group with a fixed small number of scientists. Even though the traditional ideal of a scientist is that of a lonely man working in his lab or at his desk, the notion of science being a community effort is well established in philosophy of science. What is relatively new to science and has so far not been examined by philosophers as much is what has been called "radically collaborative research" (Huebner et al., 2017; Kukla, 2012; Winsberg et al., 2014), where the research is shared between many different research groups.[23] In climate science the research is often not only scattered between different research centres but between different generations of scientists.[24] No modern climate model is built directly from scratch. Climate scientist frequently rely on bits of coding or even whole model parts that were originally developed for a predecessor of the current model, often by a previous generation of scientists (Knutti et al., 2013). Thus, decisions about how to model specific processes made 'back in the days' are still present in today's models. How these past decisions influence the performance of models can be hidden in many ways within the models (Winsberg 2012). Further it is not, for the so inclined scientists, "foreseeable how methodological choices in model development will shape modeling results in the long run" (Parker and Winsberg, 2018, p. 141; for an example see Lenhard, 2018, pp. 839–840).

On top of the intricate model history, other features of climate modelling, such as the high interdependency between different model parts, the fuzzy modularity and the need for tuning, mean that ultimately the model can only

22 This argument was first made by Winsberg (2012), see also Parker and Winsberg (2017) and Winsberg (2018, pp. 130–153).

23 Besides climate science, examples for this kind of fractured research can be found in modern physics amongst other research fields. One might, for instance, think of medical research (Kukla, 2012) or modern astrophysics. For an example from the field of gravitational waves, as made by Collins (2014), see Chapter 4.2.1 and see also Collins (2017).

24 Research centres usually develop their own models although cooperation between different institutes are, of course, taking place. One such example is the new ICON modelling framework that the Max-Planck-Institute for Meteorology developed together with the German Weather Service (Zängl et al., 2015). Furthermore, models might also incorporate specific parts that were originally developed for other models (Parker, 2018).

be fully evaluated within the context of the complete model. It is, therefore, often difficult to say whether adding a new element to the model delivers the desired results because of the new component on its own or is not also the outcome of some compensating effect resulting from the interdependency of different elements of the model. This "epistemic opacity" or "analytic impenetrability" (Baumberger et al., 2017; Humphrey, 2004; Lenhard and Winsberg, 2010) of climate models also cannot (at least practically) be resolved by reconstructing the model completely from the ground up, as Winsberg points out:

> Of course the modeller could – in principle – rework the entire code. The point is, however, that in even moderately complex cases, this is not a viable option for practical reasons. At best, this would be far too tedious and time-consuming. At worst, we would not even know how to proceed. (Winsberg, 2018, p. 143)

In this context it is evidential why any claim similar to that of Jeffrey's that the scientist, instead of making value judgements, should factor the uncertainty estimates concerning different methodological options into their overall assessment cannot be maintained in the context of complex computer simulations. As Winsberg (2012) notes, climate scientists (and scientists more broadly (Steel, 2015)) cannot be viewed as perfect Bayesian actors.[25] The climate system and the models that scientists work with are too complex for any one person to have a complete understanding of the effect of all trade-offs and prioritising on the model.[26] Put in a different way, the high complexity of climate models with its hundred thousands of lines of code, decade long construction history,

---

25 Winsberg (2018, 2012) points out that, contrary to Douglas' (2000) claim that the need for inductive risk assessment in the context of methodological decision making, shows the inevitably of social value judgements in science, this does not, in and of itself, refute the assertion that scientist cannot avoid social value judgements. One might still claim that the scientists only have to factor all issues with a particular methodological choice into a probability assessment, which they can pass on as an expert judgement to stakeholders, provided that we assume a classical Bayesian understanding of probabilities and the actors holding them (see Chapter 3.1.2). Winsberg discusses this as the "Bayesian response to Rudnerian and Douglasian arguments from inductive risk" (BRAIR). For a more formal discussion of the Bayesian argument, see Steel (2015) and why it fails in the context of social values in climate science, see Parker & Winsberg (2018) and Winsberg (2018).

26 In practice, a insufficient documentation of different modelling steps can make the models even more obscure (e.g., tuning, see Chapter 2.1).

its "fuzzy modularity" (Lenhard and Winsberg, 2010) and ever further development make it hard to imagine where scientists should even begin to disclose the whole reasoning process behind every decision ever made in the model construction process (Winsberg, 2018, p. 143). The labyrinthine way of the "nooks and crannies" (Winsberg, 2012, p. 132) of climate modelling makes it impossible to fully trace the consequences of all possible social value judgements made in the model building process.

To be clear, the crux here is not that every epistemic gap is necessarily filed by social values but that, when reviewing a whole climate model, it is not possible to evaluate retroactively to what extent and at what point exactly what kind of social values were a relevant factor in a decision-making process, as well as in what way and if at all they have influenced today's model outputs.

3.1.3.3 Coarser uncertainty quantification and other possible counterarguments

Originally, the discussion about the influence of social values in climate science mostly centred on uncertainty quantification derived from MIPs. In this context there has specifically been some disagreement about how severe the influence of social values actually is. While Winsberg (2018, 2012, 2010) has argued for the possibility of social values filling the void left by epistemically unforced decisions in climate modelling, which cannot be (fully) accounted for during the evaluation process of models, others (Parker, 2014; Schmidt and Sherwood, 2015) have argued that this problem would be at least significantly reduced when scientists are not forced to give a precise estimates of uncertainties but rather are given the option to express uncertainty in ranges of probability. Such a more coarsely grained scheme for uncertainty quantification is what is used by the IPCC for scientists to express their degree of certainty in assessing the current state of climate research. In the *Guidance Notes for Lead Authors of the IPCC Fifth Assessment Report on Consistent Treatment of Uncertainties* (Mastrandrea et al., 2010) authors of the IPCC assessment report are given a guideline which they can and should refer to when conveying uncertainties (see Figure 2).[27] Instead of requiring the scientists to express uncertainty estimates in fixed and precise numbers, they are given wider intervals. The issue of social values in climate modelling can be mitigated this way, argues Wendy Parker:

27 A further discussion of the *Guidance Notes for Lead Authors* will follow in Chapter 3.3.3.3.2.

even if social values sometimes do come into play in the model development process in the ways suggested by Winsberg, the influence of those values on estimates of uncertainty will be reduced when coarser estimates are given. The influence will be reduced insofar as choices in model development will less often make a difference to the uncertainty estimates produced. (Parker, 2014, p. 28)

Figure 2: Likelihood scale from the Guidance Note for Lead Authors for the 5th assessment report.

Table 1. Likelihood Scale

Term*	Likelihood of the Outcome
Virtually certain	99-100% probability
Very likely	90-100% probability
Likely	66-100% probability
About as likely as not	33 to 66% probability
Unlikely	0-33% probability
Very unlikely	0-10% probability
Exceptionally unlikely	0-1% probability

* Additional terms that were used in limited circumstances in the AR4 (*extremely likely* – 95-100% probability, *more likely than not* – >50-100% probability, and *extremely unlikely* – 0-5% probability) may also be used in the AR5 when appropriate.

Source: Mastrandrea et.al., 2010, p. 3, Table 1

At first glance this might be a satisfying solution, but a closer look reveals that it does not solve the underlying problem. It is likely that a coarser scale for communicating uncertainties will reduce the influence of social values. But the complexity of climate modelling prevents us from evaluating how well this actually works. It is for the same reason that it is not feasible for scientists to keep track of every possible way in which values might have an influence on the model-developing process and then communicate resulting uncertainties to policymakers and the public. The number of epistemically not fully determined decisions are so large and stretched over such a big timeframe that it seems unimaginable how it should be possible to monitor or quantify the deci-

sion-making process in all its details. Nor would it be a viable solution to advice scientists to go on to develop *every possible* variety of a model and then evaluate the impact of every single decision made on the performance of the model with statistical methods. As it is usually a multiple year-long project to build a climate model of the scale of an ESM, it is practically not possible to build every thinkable versions of a model. Climate scientists do use *model intercomparison projects* (MIPs) to explore structural uncertainties, but MIPs are not statistical evaluation methods in the sense of, for instance, a Monte-Carlo study (see Chapter 3.3.3.3). They do not sample from the whole space of all models. Instead, they are better described as "ensembles of opportunities"[28] (e.g., Parker, 2010, pp. 270). Parker and Winsberg note in this context that climate models are used precisely "because it can be very difficult to reason about such systems without them" (2018, p. 140). That is, when reasoning about the effects of some modelling decision it often requires scientists to make use of their previous experience with the available models. However this makes it questionable "how 'escapable' the model-based influence of nonepistemic values" (Winsberg and Parker, 2018, p. 140) is even if one resorts to giving coarser estimates.

What is more, the option of a more coarsely grained uncertainty scales brings with it its own new sort of value judgements. Scientists now have to make a decision which kind of scale to use. And the underlying assumptions might well be influenced by social values, as Winsberg argues:

> it seems clear that at least sometimes it is a consideration of the likely applications of an uncertainty report that guide the choice between a wider and more confident report and a narrower and somewhat less confident report. Perhaps a narrower, even somewhat less confident interval is thought to be more useful for policy makers. In such cases, social values are once again playing a role. (Winsberg, 2018, p. 149)

On the other hand, one might also question whether the gaps left by methodologically unforced decisions are necessarily filled by social values. Parker argues that oftentimes pragmatic considerations instead of social values are the decisive factor:

> Suppose a group of climate scientists is further developing their climate model now that more computing power is available. Which physical process

---

28 MIPs will be further discussed in Chapter 3.3.3.3.

> should they "add" to their model next? Suppose their choice is epistemically unforced, i.e. they cannot argue on purely epistemic grounds that one process in particular should be added next. Must their choice then either be arbitrary or determined by social values? No. Pragmatic factors can also fill the gap. For instance, the scientists might already have in hand some computer code for process P but not for processes Q, R, or S. Or they might judge that it will be much easier to incorporate P than to incorporate Q or R or S, given past choices in model building. Or they might be experts on P but have much less understanding of Q and R and S. Or it might be that a leading modeling group incorporated P for reasons like those just identified, and now it is seen as de rigueur for state-of-the-art climate models to include P. And so on. Indeed, it is plausible that pragmatic factors like these often influence or even determine model development choices. (Parker, 2014, p. 27)

It is quite plausible that pragmatic considerations can be significant in climate model building. But is that sufficient to completely rule out the potential influence of social values in climate modelling? Clearly not.

First of all, one might ask what "pragmatic" in this context even means. What exactly differentiates social from pragmatic values? And are social and pragmatic necessarily mutually exclusive? Do not pragmatic and social considerations sometimes overlap? One might very well imagine a situation where the decision to fall back on a pragmatic choice such as relying on a model part that is already well known to some of the scientists involved might be influenced by underlying social values. Those may be that a timely solution is valued more even, if it is at the expense of a possibly more precise or detailed answer, because climate change is an issue requiring urgent answers from science. More generally speaking, the decision to go with a pragmatic solution is always based on the (possibly social) value-laden decisions that a pragmatic approach is justified in this situation, as Anna Leuschner rightly notes (2016, p. 79). Furthermore, Leuschner argues, even when a certain piece of code is chosen on the basis of social-value-free pragmatic factors, it is still quite possible that the development of that model component or even just a part of the model component has been influenced by non-epistemic assumptions at one point or another.

Here again it becomes apparent how deep within the "nooks and crannies" (Winsberg, 2012, p. 130) of complex computer modelling value judgements can lie. Pragmatic factors or not, the questions remain to what extent it is actually

possible to mitigate the influence of social values and how one would even go about measuring this. Retroactively, there is neither a way of knowing for certain how much influence social values had on a climate model nor is it in any way possible to keep track of every thought that went into the construction of a model. What is so significant is not the particular impact of social values in climate modelling but that we have no way of fully tracing their influence.

But if we cannot retrace and evaluate the influence of social values, it seems also not feasible to restrict exactly what kind of social values should be allowed to impact the scientific process. Kirsten Intemann (2015) proposes that social values can be considered adequate in climate modelling as long as they are democratically supported beyond science. But it is not clear how this could actually be put into practice, let alone be monitored, even if one considers a less complex field of science than climate science. It should be noted that Intemann does not expect scientists to consult the general public at every step of the way. She argues that allowances should be made for a certain flexibility how stakeholders are chosen and to what extent their values actually should match those of whom they represent so that "modeling decisions can be more or less justified in degrees depending on the extent to which social and epistemological aims are clear and there is evidence that they would be broadly endorsed" (Intemann, 2015, p. 228). Further, she argues that the scientist is not required to consult the stakeholders at every step of the way, rather scientist and stakeholder are in "a process of interactive feedback loops" (Intemann, 2015, p. 288). Still, there is no way of knowing if just democratically determined values (even if only to varying degrees) are really the only ones that play a role in the decision-making process in climate modelling. On the contrary, it seems very questionable to me that they are all necessarily democratically supported. The sheer number of methodological underdetermined decisions, again, makes that very unlikely. After all, scientists are not a homogenous group, so we can expect them to have a variety of values and, as epistemically not fully constrained choices are a consistent feature of climate modelling, it is questionable whether they actually check regularly if their choices are in accordance with the wider societies' values. But whatever might determine those "thousands of methodologically unforced decisions" (Winsberg, 2012, p. 130), we might never entirely know for at least a significantly large number of them.

All this might lead one to question why some philosophers and even climate scientists[29] are so determined to show that the influence of social values in climate modelling is either negligible or can somehow be (democratically) legitimised. I think Winsberg is right when he argues that at the root of this is a misunderstanding and that value-ladenness "is not at all the same as the claim that scientific conclusions are reached in a way that is systematically biased" (Winsberg, 2018, p. 150). Mere value-ladenness is not does not mean outright bias.

3.1.3.4 Systematic bias and wishful thinking

This leads us back to the main concern of the proponents of the value-free ideal: social values in science will inevitably lead to bias and wishful thinking. At first glance, this might seem to be a reasonable concern. After all, climate science seems to be saturated with all sorts of possible social-value-type assumptions. And as climate change is a highly political topic, one might assume that this could give scientist plenty of opportunities to influence the models consciously or unconsciously in a way that suits them best. However, a closer look at the situation shows that this concern is unwarranted.

First of all, the number of scientists involved in the development of a global climate model (at least in the case of those of the complexity of an AOGCM or ESM) make it highly unlikely for it to be possible for one scientist to single-handedly 'sabotage' a model, at least not without being noticed by their colleagues. However, a follow-up claim might then be that climate scientist collectively consciously or unconsciously influence the model developing process in a way so the models are in accordance with their personal social convictions. In the first case, we would imply that there is a grand conspiracy at play involving huge parts of the climate science community. If we disregard this rather outlandish assumption, the second possibility is a little bit more complicate to refute. This claim relies on the assumption that scientists are a monolithic group in one way or another. Looking back at the history of public perception of climate science it has not been an uncommon occurrence for climate-change deniers to accuse scientists of being biased. Either because they (unconsciously) fear that their work would otherwise be redundant or because they cannot separate their work and their personal political convictions, as described in the introduction to this book.

29 For examples of the former, see also Betz (2013) and for the latter, see Schmidt and Sherwood (2015).

The worry that the value-laden background assumptions of scientists can have an inappropriate effect on scientific research in general is also voiced by many feminist philosophers of science. Longino (2002, 1990), for instance, advocates for a pluralism of perspectives in science. The hope is not that, within a diverse group of scientists with different backgrounds (social, racial, political, and so forth) compared to a group of scientists with a homogeneous set of values, the influence of social values can be prevented but that their impact will be easier to detect. Leuschner (2012a) argues that the IPCC's structure, regarding the selection of scientists, is similar to such a kind of pluralism.[30] As an intergovernmental organisation the IPPC specifically selects the authors "taking into account the range of scientific, technical and socio-economic views and backgrounds, as well as geographical and gender balance" (IPCC, 2023). This, argues Leuschner (2012b, pp. 176–177), fulfils two purposes: on the one hand, the hope is that all involved countries feel included and are, therefore, more inclined to implement mitigating climate policies. But on the other hand, there is also the epistemic expectation that this will ensure that all relevant knowledge and data is taken into account under the assumption that scientist have unique and specific scientific, cultural and political knowledge about the region or country they come from. The IPCC itself also states that the reason for diversifying the field of participating scientists is "to ensure that reports are not biased towards the perspective of any one country or group of countries and that questions of importance to particular regions are not overlooked" (IPCC, 2023). The purposeful inclusion of minorities in science is not an unscientific act fuelled by social values but rather follows good epistemic consideration.

30 To be more precise: Leuschner agrees with Longino on the necessity of a pluralist approach to science in order to reveal hidden value assumptions, but she also criticizes that Longino's idea how this should be implemented in practice "suffers from an inherent circularity" (Leuschner, 2012a, p. 197). Longino argues for a pluralism that simultaneously demands that everybody ought to be able to participate in the critique of scientific discourse but also to exclude any unqualified opinions. This is contradictory and circular, notes Leuschner, as it is not clear how any kind of standards defining what qualifies contributors are is to be determined, without constricting the pluralistic process of including as many perspectives as possible. Inspired by Kitcher's concept of "deliberators" (2001), Leuschner argues for a pragmatic and situation-specific solution for this problem including pluralistically organised but politically installed expert groups, which would evaluate scientific practices and findings, such as it is the case with the IPCC.

However, it is not just the outer socio-scientific structure of organisations like the IPCC that implement procedures to constrain bias, but also the complexity of both the climate system and the models in and of itself that helps prevent wishful thinking affecting research. Complex climate models of the scale of an ESM or AOGCM are used for multiple purposes. Contrary to the public perception, the main goal of climate science is not just to further pin down exactly how the climate will change under a certain emissions scenario but rather advance the understanding of specific climate processes and the climate system as a whole. Considering the amount of money and effort going into developing a new model of this type, it seems obvious that scientists and financial backers are very much interested in developing models that can be used for a variety of purposes.[31] However, this also means that different researchers or research groups that are involved in the development process come with slightly different agendas to the table.

What is more, the argument that climate scientists could just adjust the models to their own preference loses in strength when one considers that various parameters of interest do not exist in isolation in the models. Quite often changing one parameter also directly influences other. Thus, as not all variables and processes that the scientists would like to explore with the model can be equally well represented, scientists are again confronted with having to make trade-offs with respect to their competing preferences. Here again is a kind of pluralism at play.[32]

Climate science, thus, gives us a perfect example why possible value-laden deliberations within the inner-scientific process are by far not as much of a threat to science as the discussion about them in philosophy of science lets us believe. Not because they are so rare or obvious to spot. On the contrary, there is a myriad of ways in which social values might influence scientific processes, yet because of that they are not just unavoidable but also mostly epistemically harmless. Additionally, as this chapter has shown, fuzzy modular-

31 There is currently even a trend to develop models that can simultaneously be used for climate modelling purposes as well as weather prediction. This was first done with the Unified Model of the Met Office in the UK. In Germany, the Max-Planck-Institute for Meteorology and the Deutscher Wetterdienst (German Weather Service) have also joint forces to develop a shared model framework (ICON).

32 Besides that, one has to remember that climate scientists rely on more than ESM to assess the impact of climate change, such as a variety of global models, regional models and empirical data from observation and experiments from different fields of research and expert judgement (see Chapter 3.3.3.4).

ity, compensating effects and entailing trade-offs mean that it is simply impossible to tune the 'perfect' model. The complexity of the climate modelling is simultaneously the reason why value-laden considerations may enter the climate model building process as well as the very feature that protects climate science from unwanted (conscious or unconscious) political or social influence. The consequences of every modelling decision can be so manifold and inscrutable that it makes it in fact much harder to argue that it is even possible for scientists to influence the models effectively in a way that suits their own social or political beliefs (Parker and Winsberg, 2018). Further, representational risks, which are likely a bigger source of social value-laden decision making in climate modelling than inductive risks do not constitute "influences that make it more likely that one conclusion rather than another will be reached" (Winsberg, 2018, pp. 150).

3.1.4 Conclusion

Nowadays the vast majority of philosophers of science accept that value judgements are an unavoidable element of science. But there is a lively discussion ongoing about what exactly the appropriate role of social values is and how the non-epistemic, social realm can be constricted. While Intemann, for instance, argues that "value judgments are legitimate when they promote democratically endorsed epistemological and social aims of research" (2015, p. 217), Douglas (2009) suggests that social values at stages internal to science should be restricted to an indirect role. Both positions have in common that they argue that the influence of social values must be restricted and limited to specific cases one way or another.

When it comes to a differentiation between the direct and indirect role of values, it seems questionable if that distinction is particularly helpful here. The distinction that Douglas makes is fuzzy to begin with. Douglas notes that there are exemptions for both kinds of roles of values. Thus, values might play a direct role in the inner-scientific process such as when the ethical implications of methodologies have to be unexpectedly reassessed. In the same way, according to Douglas, social values might also be inappropriate in the pre-scientific context when they undermine the core scientific goal of gaining knowledge.

Further, inductive-risks assessment is not the only way in which values play an unavoidable role in science-internal processes. As Harvard and Winsberg (2022) note, when it comes to representational risks, determining the appropriate role for social values by distinguishing between a direct and indirect role

of social values is no longer a viable route.[33] Predictive preferences and cost deliberations interfere in a much more direct form. Note that these kinds of value judgements are not just decisions under uncertainty but, in fact, decisions of scientific research objectives.

It has to be emphasised here that predictive preferences are not necessarily priorities which are set before the model is developed. There might be certain priorities that scientists define before setting out to create a model. But these have to be rather general as the model cannot be planned in all its details from the beginning. Climate modelling involves a certain degree of tinkering and trying out which method works best for the specific model (Held, 2005).[34] The complexity of these kinds of models makes it impossible to anticipate every decision necessary in its construction. Thus, the research goals and priorities are constantly under some threat of having to be reset and readjusted (to some degree at least). Further, the number of people involved, often over several generations, makes it unlikely that they all share and abide by the same interpretation of these priorities. This is further complicated by the fact that by relying on model parts, whether whole parametrisations schemes or bits of code originally developed for different models (in other words by not constructing the model from scratch) choices made decades ago will constrict the modelling process. Therefore, one cannot simply view these *predictive preferences* as pre-scientific goal setting, that is the kind of social-value interference that even most proponents of the value-free ideal see as unproblematic (Douglas, 2009, p. 45).

The value-free ideal has always been an illusion, created by science to protect itself against unwanted interference from religion (Rudner, 1953) or for fear of losing its authority (Douglas, 2009, p. 79). As much as the supporters of the value-free ideal have tried to deny it, values have always been part of science. Science is neither all of a sudden overrun by social values nor has science become unreliable and biased. Such a view of science overlooks the fact that this is a state that science has always been in. The increase of complex systems as the subject of research in science just makes the illusion of the value-freeness of science that much more obvious.

---

33 Although the authors come to the conclusion that where the risk of endorsing as false fact is concerned, limiting social value judgments to an indirect role at an internal scientific stage might still be a good way to rule out wishful thinking (Harvard and Winsberg, 2022).

34 See also Chapter 4.2.2.

It is, of course, quite understandable why some climate scientists have initially reacted hostile to the suggestion that social values might be at play in climate modelling (e.g., Schmidt and Sherwood, 2015).[35] After all, they are regularly under fire from climate-change deniers, who accuse climate scientists of being misled by their own personal convictions. As Proctor puts it:

> Value-freedom is an ideology of science under siege – a defensive reaction to threats to the autonomy of science from political tyrants, religious zealots, secular moralists, government bureaucrats, methodological imperialists, or industrial pragmatists asking that science be servile or righteous or politically correct or practical or profitable. (Proctor, 1991, p. 68)

Climate science has very much been such a "science under siege". In such a situation, it might be tempting (and often initially successful) to insist on the value-freeness of one's own research, but scientists do science as a whole a disservice when they keep insisting on practicing value-free science. Asserting that the value-free ideal is still upheld – born out of an instance of self-defence – chances are high that it will backfire in the long run because the more the complexity of the systems that scientists investigate increases, the less likely it will be that scientists can successfully hide behind an apparent value-freeness. Science – as most human enterprises – cannot, has never and will never be value-free.

While there might have been originally some scepticism from within the climate-science community when philosophers began to discuss the role of social values in climate modelling, it also has to be noted that the role of social values in climate science are actually openly discussed in the latest IPCC report. Particularly, predictive preferences (though not named as such) being an unavoidable element of climate modelling are highlighted:

> Social values are implicit in many choices made during the construction, assessment and communication of climate science information (Heymann et al., 2017a; Skelton et al., 2017). Some climate science questions are prioritized for investigation, or given a specific framing or context, because of their

35 Gundersen has shown in a small study of a group of Norwegian climate scientists that many of them stand by the value-free ideal, although the scientists also note that it is sometimes difficult to guarantee the value-freeness in practice. Gundersen remarks as well that some scientists observe that "strict adherence to the value-free ideal can undermine policymakers' perception of the relevance of experts' opinions" (2020, p. 113) when it makes them to be too cautious in conveying the significance of their findings.

> relevance to climate policy and governance. One example is the question of how the effects of a 1.5°C global warming would differ from those of a 2°C warming. [...] Likewise, particular metrics are sometimes prioritized in climate model improvement efforts because of their practical relevance for specific economic sectors or stakeholders. [...] Sectors or groups whose interests do not influence research and modelling priorities may thus receive less information in support of their climate-related decisions. (Chen et al., 2021, p. 172)

This quote also points us toward another important distinction that needs to be made. I have argued above that value judgements in climate science are in most cases epistemically harmless; however, that does not mean that there are not at the same time some non-epistemic risks that can arise out of predictive preferences, to the extent that, e.g., a lack of attention towards the predictive preferences of underprivileged community can cause harm when their need for a particular kind of knowledge is not taken into account (Harvard and Winsberg, 2022; Parker and Winsberg, 2018). One way to mitigate this risk seems to be (again) a pluralistically organised scientific community (Jebeile and Crucifix, 2021).

So far the discussion of social values in science has often centred on the argument from inductive-risks. But, as has been shown in case of climate science, representational risks are also an unavoidable part of developing a complex computer simulation. Further, we have also seen that with the increasing complexity of science it becomes more and more impossible to retroactively make those value decisions explicit. As I have argued above, this should not be seen as an epistemic problem. Instead value judgements ought to be regarded as a necessary part of science. Particularly when it comes to complex computer simulations, they fill gaps left by epistemic and methodological underdetermination. While the complexity of the system introduces an inability to trace the effects of social-value deliberations through the model building processes, it also works simultaneously as a safeguard against the directed value-laden manipulation of the models or, in other words, the complexity 'inoculates' the models against wrongful influence of this kind. On the one hand, the specific values are so numerous and diverse and quite often do not even have the "right form" (Winsberg, 2018, p. 151) to make the models biased in a specific way. On the other hand, the number of scientists involved works as an insulation and corrective tool against individual bad work. Particularly, when the group of modellers is sufficiently diverse, it makes it more likely that "later choices in

model development can 'undo' the effects of earlier ones" (Parker and Winsberg, 2018, p. 141, see also Jebeile and Crucifix, 2021). Whatever may be the case, the issue is not whether or not the scientist's decisions are directly guided by social consideration, but that we can no longer retroactively tell whether that has been the case or not (Winsberg, 2012).

Under these circumstances it is also questionable if a distinction between cognitive and social values is still useful to determine the appropriate role for value judgments in science. Even if we were certain that all modelling decisions were determined by cognitive values alone, one still cannot be sure that these are not affected by the specific social context under which they were constructed. Considering Longino's (2008) claim that we can just as well imagine alternative set of cognitive values, which are as well justified as the traditional ones, one has to at least question if in a similar way specific modelling cultures might be in a hidden way informed by social or ethical background assumptions (see Chapter 3.1.1.2). So it does not even make sense in this context to discriminate between some kind of science-internal, appropriate and extra-scientific values that are only under very specific circumstances allowed to interfere with scientific processes.

Though the fear of the influence of social values on science is historically understandable, we need to change our perspective on values in science. Against the backdrop of the vast number of epistemically not fully constrained decisions and the new epistemic challenges of dealing with highly complex systems, the discussion about values in science must shift from a discussion of what the appropriate role of values in science is to what an inappropriate role would be. Douglas has argued that "values should never suppress evidence, or cause the outright rejection (or acceptance) of a view regardless of evidence" (2009, p. 113). Despite the vast variety of necessary roles that values judgements can assume in the construction and evaluation of complex computer simulations, this also seems to be a prudent approach for inductive risks. However, predictive preferences, which seem to be the biggest source of possible social value-laden assumptions, do not hold a clear risk for scientists to outright disregard evidence as it is primarily a question of research objectives (see Harvard and Winsberg, 2022). Here the bigger, non-epistemic concern is that some underrepresented stakeholders might get less information concerning their particular circumstances. Nevertheless, as has been argued above, from a purely epistemic perspective, the risk of inappropriately influencing climate-

model construction through wishful thinking or deliberate bias does not seem to be a particular high risk.[36]

3.2 Model, theory and observation

3.2.1 Introduction: from handmaiden to a life of their own

For a long time the experimental and practical part of science has been somewhat neglected by philosophy of science. Until well into the middle of the 20th century the discourse in philosophy of science has focused primarily on a discussion of scientific theories. While Francis Bacon saw the experiment at the centre of the scientific enterprise, by the time philosophy of science had become a discipline of philosophy in its own rights, at the beginning of the 20th century, observations and experiments had been cast into the role of the "handmaiden of theory" (Gooding, 2000, p. 119). Meaning that the primary purpose of the empirical part of science was seen as to provide data to evaluate theories. According to this theory-focussed view of science, experiments and observations are only of relevance to science once a theory has been developed and needs to be tested. They were considered to be of little philosophical interest on their own.

This disregard of the experimental and observational element of science in the history of philosophy of science becomes most apparent in the approach that logical empiricism and critical rationalism take to this issue. While the logical empiricists focused on the logical and theoretical foundation of science, they reduced the empirical part of science to producing basic observational sentences (*Beobachtungssätze*).

This indifference towards the practical part of science in the first half of the 20th century did not just hold for logical empiricism. Karl Popper, who in

36 In effect, cases where climate-change deniers have argued against mainstream science qualify as exactly one of those situations where the deniers have neglected the evidence in favour of personal beliefs. Naomi Oreskes and Eric M. Conway (2010) show compellingly how a small subset of scientists, often paid by specific interest-groups, such as oil companies, have disputed a variety of scientific claims from smoking causing cancer to climate change over many decades because of their dislike of governmental regulations.

many ways opposed the philosophy of the logical empiricism, still saw the experimenter as someone who does the legwork while being fully guided by the theoretician:

> The theoretician puts certain definite questions to the experimenter, and the latter, by his experiments, tries to elicit a decisive answer to the questions, and to no others. [...] But it is a mistake to suppose that the experimenter proceeds in this way 'in order to lighten the task of the theoretician', or perhaps in order to furnish the theoretician with a basic for inductive generalizations. On the contrary, the theoretician must long before have done his work, or at least what is the most important part of his work: he must have formulated his question as sharply as possible. Thus it is he who shows the experimenter the way. But even the experimenter is not in the main engaged in making exact observation; his work, too, is largely of theoretical kind. Theory dominates the experimental work from its initial planning up to the finishing touches in the laboratory. (Popper, [1935] 1959, p. 107)

Popper does not consider the work of the experimenters as completely superfluous, but he also does not see them as the ones who take initiative. The experimenter, according to Popper, is not in a position where they could contribute anything substantially new to science on their own. On the contrary, the job of the experimenter is seen here as only to confirm or falsify hypotheses as instructed by the theoretician. Thus, to Popper the experiment takes a subordinate role to the theory. It provides evidence but cannot in itself provide new scientific insight. It is for this reason that Popper gives the *theory* that much more attention than the *experiment* in his writings.

Just as in the case of the rise of the value-free ideal this development coincides with the rising popularity of the distinction of the *context of justification* from the *context of discovery* made by Reichenbach (1938). As already discussed in more detail in Chapter 2.2, this distinction became very popular among philosophers of science of the 20th century, when it came to separating the realm of philosophy of science from that of psychology and social studies. Proponents of this view are of the opinion that philosophy of science should focus on logical justification of a scientific discovery, not the practical path leading to it. The question of how a scientific fact, theory or law is discovered is thereby made a matter of sociology or psychology but not philosophy. In that respect the DJ distinction also played an important role in directing the attention of philosophers of science towards theories (Schickore and Steinle, 2006a). Experiments

and observations were seen to be only tangentially philosophically important insofar as experiments are the way to provide empirical evidence for theories. However, the process through which the necessary data for this is acquired was seen as by and large not appealing to philosophical contemplations. The experimental part of science was, thus, (dis)regarded as (for the most part) an element of the context of discovery and, thereby, cast aside as philosophically uninteresting.

For most of the 20th century the default position seemed to be to omit the experimental part of the scientific enterprise from the philosophical discourse, or as Gooding put it: "Experiment seems to be an epistemological football – essential to the game, but of no intrinsic philosophical interest" (Gooding, 2000, p. 122). However, beginning in the late 1970s, philosophers and sociologists of science increasingly started to question the theory-dominant view of science and set out to bring "studying scientific practice, what scientists actually do" (Pickering, 1992, p. 2) back into the limelight. During what is today sometimes described as the *practical turn* or the *new experimentalism* they began to explore the different functions and characteristics of experiments and other aspects of actual scientific practice (e.g., Ian Hacking, David C. Gooding, Allan Franklin, Nancy Cartwright). Though they all highlighted the necessity to include experimental scientific practice in philosophical discussion about science, there is a certain disagreement to what extent experiments can be considered independent from theory (Feest and Steinle, 2016; Gooding, 2000).

Ian Hacking for instance argues that experimentation can happen independently from theories. By means of a number of examples from the history of physics, Hacking shows in his well-known book *Representing and Intervening* how varied the dynamic between theory and experiment in actual scientific practice can be (1983, pp. 149–165). While sometimes the theory preceded the experiment, quite often experiments were done independently of a specific theory. One example Hacking gives of such a case is the early days of optics where experiments were done without any fully established theory. Another case concerns the discovery of cosmic background radiation which was discovered experimentally independently of a corresponding theory that was developed elsewhere at the same time by different scientists. For Hacking experiments are as philosophically intriguing and important in knowledge generation as theories. Further, the experimentalist does not rely on the theorist

to provide them with a hypothesis to (dis-)confirm.[37] The experimental part of science has, as Hacking puts it, "many lives of its own" (1983, p. 165).

Despite these more recent trends, there is still quantitatively more literature in philosophy of science about the theoretical part of science than about the practical, empirical part. Something which can also be observed in the context of philosophy of climate science. While much has been written and said about the use of computer simulation in climate science, less can be said so about the creation and evaluation of observational data.[38]

In the following I will argue that, in fact, the process of empirical observation making in climate science is just as philosophically interesting as the computer simulations used to model the climate system. Furthermore, a closer look at the complex process of the production of climate data and the intricate relationship between observational data and climate models will reveal that conventional ideals about the role of observations in science, similar to those expressed in the quote by Popper above, cannot be maintained. As will be shown, making observations and constructing models are neither fully separated processes nor can it be said that observations provide irrefutable benchmarks to distinguish good models from bad ones. Particularly, a widely discussed controversy about satellite data will show how a widespread presence of these idealisations of scientific procedures in the public's understanding of science can be capitalised on by climate-change sceptics and so inclined interest groups to sow doubt about the trustworthiness of climate science.

I will begin in Chapter 3.2.2 with a short general philosophical debate of some relevant philosophical concepts, particularly *theory-ladenness* and *models of data*, before I turn to the specific case of observational data in climate science. But before doing so, it seems prudent to first consider the definition of the term *observation*.

37 All this has, of course, to be understood in the context of Hacking's *entity realism*, according to which experiments can confirm the existence of entities independently from theories. As he famously puts it in respect to the use of electrons in experiments: "if you can spray them, they are real" (Hacking, 1983, p. 24).

38 Some noteworthy exemptions are Edwards (2010); Guillemot (2017, 2010); Lloyd (2012); Parker (2020, 2017). Edwards (2010) specifically writes an extensive historical account of the development of a meteorological and climatological infrastructure.

3.2.1.2 Observation

In the context of the empirical part of climate science, scientists usually describe their work as making and processing *observations*. But what actually defines *observations* and what makes them distinct from *experiments*? And to what extent can this distinction be made at all? To answer this question let us first take a look at a short history of the term, as recounted by Daston (2011) and Daston and Lunbeck (2011).

From a historian's perspective, the use and meaning of the terms *observation* and *experiment* in science itself has changed more than once over the last four centuries from synonyms to antonyms, as Daston explains:

> In the period from the early seventeenth to the mid-nineteenth century, the relationship between observation and experiment shifted not once, but several times: from rough synonyms, as in the phrase "observations and experiments" that had become current in the early seventeenth century, to complementary and interlocking parts of a single method of inquiry throughout much of the eighteenth and early nineteenth century, to distinct procedures opposed as "passive observation" and "active experiment" by the mid-nineteenth century. (Daston, 2011, p. 82)

While empirical science as such becomes gradually more relevant in the late 17th century, the term *experiment* becomes narrower and now refers to "deliberate manipulation" (Daston, 2011, p. 85) or what Bacon called *artificial experiment*; the term *observation* becomes wider. Observation making, disregarded by medieval scholars "with conjecture because its results were uncertain" (Daston, 2011, p. 104), had become an activity so relevant to science by the middle of the 18th century that it had become "a way of life" for many in the scientific community, dictating their daily routine (Daston, 2011, pp. 101–104).

Towards the end of the century, notes Daston (2011), observations had become a full-fledged 'instrument' of thinking and reasoning in its own right, including repetition of observations and comparison to others. Observation making as a whole had taken a distinct, methodological, systematic and communal character. It had become something which most scientists saw central to their work and reasoning processes.

In the 19th century the terminology underwent a new shift, which singled out the experiment as the activity requiring real talent and training, contrary to observations:

> starting in the 1820s, prominent scientific writers began to oppose observations to experiment, and to vaunt the prestige of the latter over the former. In this new scheme of things, experiment was active and observation was passive: whereas experiment demanded ideas and ingenuity of the part of a creative researcher, observation was reconceived as the mere registration of data, which could, some claimed, be safely left to untrained assistants. (Daston and Lunbeck, 2011, p. 3)

Seeing observation making as a passive exercise that could also be done by the untrained was not a degradation of the relevance of observation to science but interpreted as an advantage to science. By being able to outsource observation making to 'untrained forces', scientists hoped to make sure that the data remained 'objective' and unspoiled by the scientist's theories. This attitude is also reflected in 20th century philosophy of science, according to Daston and Lunbeck. The attempt of the philosophers of logical empiricism to create a scientific system in which any theory can be retraced to observational protocol sentences "would render observation in a language as close as possible to the raw data of perception" (Daston and Lunbeck, 2011, p. 5).

These days two different but at the same time overlapping definitions of the two terms are very common. Today scientists usually loosely ascribe the term *observation* to data collection in a fixed target system, whereas experiments in science traditionally include manipulations of the target system. Both can be understood as empirical, scientific practices from different ends of a spectrum. While experimentation demands active intervention by the scientists, usually in a laboratory setting, making observations is seen as a much more passive activity requiring often the skilful application of measuring instruments but no interference with nature.

There is also another use of the term *observation*, that is not as widespread in science but highly prevalent in philosophy of science. This definition also sees observation as a passive activity but narrows it down even further. Here *observation* refers to the sheer perception or detection of data, quite often as part of an experiment but also as detection of natural phenomena with or without instruments.[39] This definition of the term will have specific relevance in the context of the following discussion of theory-ladenness of observations.[40]

39 For the difference in the use of the term observation see, also Shapere (1982).

40 Hacking incidentally, though dedicating a whole chapter in *Representing and Intervening* to the topic of observations, is rather imprecise in his use of the term *observation*.

One common feature of both definitions is that they both seem to have retained the notion of observation as a passive and experiments as an active undertaking. But as we will see in the following in the case of climate science, disregarding observations as just passive perception overlooks how much active knowledge of the matter at hand and training in the use of instruments are required in scientific observation making.

3.2.2 Theory-ladenness, underdetermination and models of data

The notion that theories are underdetermined by observations, meaning that there can always be an alternative theory to explain an observation, was first introduced by Pierre Duhem (1906) for theories in physics. W.V.O. Quine (1951) extended this idea to any kind of knowledge claim. Today underdetermination is often understood as the Quine-Duhem problem, as a problem of confirmational holism to the extent that every hypothesis is accompanied by auxiliary hypotheses (Stanford, 2023). That is, when a hypothesis is found not to be in accordance with the empirical observations, the hypothesis cannot be (by logic alone) deemed wrong, as it could also be the case that one of the auxiliary hypotheses is wrong.

Norwood Russell Hanson (1958, pp. 4–30) was the first to introduce the phrase of *theory-laden observations*. Hanson argues that two scientists can observe the same object but 'see' different things. Assuming a proponent of the heliocentric and one of a geocentric worldview watch the sun rise together, he poses the question: do they see different things? Hanson argues that this is in fact the case. When scientists speak of 'seeing' something, they do not refer to the mere process of physiological perception or, as Hanson put it, "seeing is an experience. [...] People, not their eyes, see" (1958, p. 6). So looking at the sun is more than the reception of photon particles on the retina of the scientist's eyes. Two

He mainly refers to observation in the context of data detection, but occasionally he gravitates towards a definition that is closer to a definition of observation as experiments with fixed target systems (1983, pp. 155–156, 180). As Malik (2017) points out, by (in effect) also defining experiments as the creation of phenomena in a pure state, Hacking omits great parts of science, which then fall neither into the category of observation, experiment nor theory, such as medical research. Hacking can only maintain this narrow definition of experiments by almost exclusively referring to examples from physics.

people having the same physical premises (e.g., both have equally well functioning human eyes) and observing the same physical object can still 'see' different things, according to Hanson. In this sense, the heliocentrist would look at the sun and see an object at the centre of our solar system, and the geocentrist sees an object circling the earth. How scientists 'see' the world and data they extract from it are, argues Hanson, influenced by their specific background assumptions.

Figure 3: Necker Cube

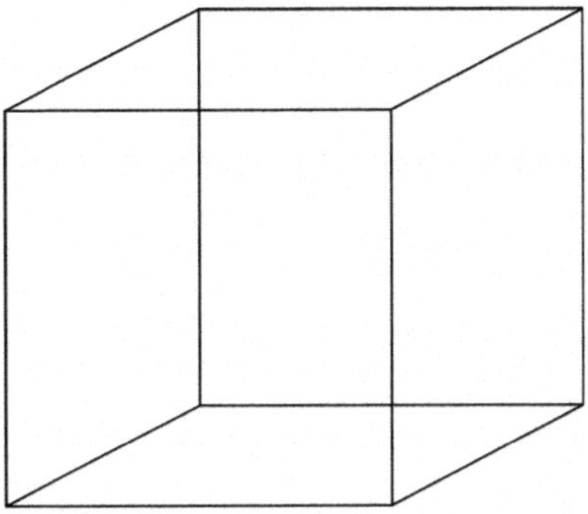

This, Hanson insists, is not just a kind of interpretation of data: "To interpret is to think, to do something; seeing is an experiential state" (Hanson, 1958, p. 11). The scientist looking at a lab sees various instruments such as specific microscopes or other special instruments. The scientist does not think to interpret the instruments as such, they just see. The layperson, on the contrary, only sees a number of cables and lenses and so forth. "The knowledge is there in the seeing and not an adjunct of it" (Hanson, 1958, p. 22). Hanson compares this to the way we perceive ambiguous images (Hanson, 1958, pp. 8–14). When we look at the Necker Cube (Figure 3), for example, some might see it as from above, others as from below. But we cannot see it in both ways at the same time. Nor would we describe the way we see the cube as an interpretation of twelve specific lines on a paper. We just *see it*; we see *a cube* from above or below, respectively. And someone who has never been taught or who does not have the physical ability to see how twelve lines of the same size, ordered in a specific way, can look like a cube will only see twelve lines. Our prior knowledge guides how and what we see, argues Hanson. It is a one-step process. Seeing and knowing

go together, insofar as the way one observes depends on the individual background assumptions and knowledge. Imagine, for instance, a physicist and a child both looking at the drawing of an X-ray tube but which to a child who has not been told otherwise just looks like a scribble of circular and straight lines (Hanson, 1958, pp. 15–19). How we observe the world is context-dependent and observations are theory-laden in the sense that an "[o]bservation of *x* is shaped by knowledge of *x*" (Hanson, 1958, p. 19).

Similar sentiments to Hanson's were also voiced not much later by other philosophers of science like Kuhn (1962) and Feyerabend (1959). It is, therefore, worth mentioning here that, although the concept of theory-ladenness has subsequently been well-established in philosophy of science, not all philosophers follow Hanson's view that all observation is necessarily theory-laden (Feest and Steinle, 2016). Hacking, for example, argues that there are instances in the history of science where scientific discoveries were made without necessarily having the right background assumptions, such as William Herschel's discovery of radiant heat (1983, pp. 167–185).[41] Hacking tells the story how Herschel, after initially noticing how different filters he had used on his telescope transmitted different amounts of heat depending on their colour, began to further experiment with a prism and a thermometer, measuring the heat of rays of light and made further measurements with all sorts of filters. He did, argues Hacking, all of this without having a fixed idea what was actually going on.[42] In the end, Herschel abandoned the experiment. But the reason why he gave up, according to Hacking, was not that he had no satisfying theory but experimental difficulties that he could not overcome. To Hacking being a good observer is much more a question of being skilled at specific observation-making practices than having background assumptions about what is observed, an aspect of observation making that I will come back to in Chapter 4.[43]

41 Hacking here applies a rather narrow concept of *theory* and defines it as "a word best reserved for some fairly specific body of speculation and propositions with a defined subject matter" (1983, p. 175).

42 Though Hacking notes that Herschel's first guess of a partially visible spectrum of light coming from the sun was close to our current understanding of what causes radiant heat (1983, pp. 176–177).

43 Hacking invokes the examples of lab technician without a university degree or the ability of William Herschel's sister Caroline to detect comets (1983, pp. 179–180).

3.2.2.1 Models of data

The idea of *models of data* has been well-established in philosophy of science since it was first introduced by Patrick Suppes (1962). Suppes notes that where the evaluation of theories is concerned one does not simply compare theories to raw observations:

> Theoretical notions are used in the theory which have no direct observable analogue in the experimental data. In addition, it is common for models of a theory to contain continuous functions or infinite sequences although the confirming data are highly discrete and finitistic in character. (Suppes, 1962, p. 253)

Instead, scientists compare *models of theory* to *models of data*. These data models are usually interpreted to be the statistical analysis of the experimental research results.[44] In this sense a model of theory is a specific "realization of the theory" and a model of data is a "possible realization of data" (Suppes, 1962, pp. 252–253).

Two examples will show how this translates into scientific practice. The first example, weather forecasting, which will also give us a first glimpse at how the notion of models of data will be relevant to climate science is given by Baas van Fraassen (2008):

> On the weather forecast website I consult I can find a graph depicting yesterday's temperature plotted against time. This was constructed from data gathered at various stations in the region, at various times during the day – this graph is a smoothed-out summary of the information that emerged from all these data, it is a *data model*. The question about the daytime temperatures in this region of one day ago is answered with a *measurement outcome*, certainly – but that is the graph in question, which is a data model constructed from an analysis of the raw data. (van Fraassen, 2008, p. 166)

Van Fraassen also emphasises that – whether a single measurement outcome or statistically processed models of data are considered – the data is also

44 Suppes specifies a model of data as "designed to incorporate all the information about the experiment which can be used in statistical tests of the adequacy of the theory" (1962, p. 258). This interpretation of models of data as statistical models can be problematic, as Leonelli (2019) points out, because this excludes certain types of data such as images.

shaped by the circumstances under which the measurements are taken. One has to look at the result as "*this is what the object looks like in this measurement set-up*" (van Fraassen, 2008, p. 167).

A similar point of view as far as the role of observation in science is concerned is expressed by Ronald Giere (2006). He argues that every observation, whether done with our own eyes or through instruments, are done from a specific perspective. That is, the use of scientific instruments does not give us a more 'objective' understanding of the world, in the sense that it provides a view from nowhere, free from the personal perspective of the scientists. There might be ways in which scientific instruments can reduce human influence and, thereby, make observation more stable, but they cannot represent the world from a universal perspective, argues Giere:

> The inescapable, even if banal, fact is that scientific instruments and theories are human creations. We simply cannot transcend our human perspective, however much some may aspire to a God's-eye view of the universe. (Giere, 2006, p. 15)

That is, instruments can only provide us with a picture of the world that is taken from a specific point of view.[45] One example of this, Giere (2006, pp. 41–49) provides, are modern telescopes as they are used by astrophysicists. First of all, there is a variety of different kind of telescopes measuring different things: radiotelescopes, gamma ray telescopes, X-ray telescopes, optical telescopes, to name only a few.[46] Furthermore, the actual physical position of the telescope is relevant: e.g., here on earth or in space. All of these telescopes would measure something differently even if we were to point them at the same part of the sky. That means, argues Giere, they show us a particular perspective of the same part of space. On the flip side, this also means that these instruments are also

45 This, of course, also holds for humans themselves. Humans have, as Giere points out, a specific (trichromatic) colour vision of the world that is the result of the interaction of our body (the perception of light rays on the retina in our eyes) with some physical processes and features of the objects (chemical setup of the object and radiation of light), which not all animals share because they are, e.g., dichromats or tetrachromats (2006, pp. 17–40).

46 Modern telescopes such as the *Hubble Space Telescope* carry instruments that can measure a wide range of wavelengths, but scientists and technician operating them must still make a decision which wavelengths are relevant for their research questions.

always 'blind' in some respects. Gamma-ray detectors in a telescope are tuned to, well, gamma rays and cannot detect radio waves.

Secondly, the data produced by these telescopes looks at first nothing like the pretty, colourful images we know from magazines or sci-fi films. Before they are useful to scientific research, this data also has to be transmitted, filtered, evaluated and corrected for background noise and measuring errors among other things.[47] All this is done by relying heavily on theoretical background assumptions. The *Hubble Space Telescope*, for instance, produces images through gravitational lensing. That means that scientists rely on the assumption that, according to the theory of general relativity, mass bends light in such a way that one can observe objects that are further away than the object the telescope is pointed at. The *Compton Gamma Ray Observatory*, by contrast, operates on the assumption that the decay of different elements releases gamma rays at specific energies, which can be detected in a rather indirect fashion by making use of the Compton scattering that the gamma rays will trigger in the detectors.[48]

Neither one of these telescopes produces the one 'right picture' nor does any of them produce a wrong one. They rather all show different perspectives of the same object, argues Giere:

> Scientific observation is always mediated by the nature of instruments through which we interact with selected aspects of reality. In this sense, scientific observation is always perspectival. (Giere, 2006, p. 43)

Depending on the instruments used and how this data is evaluated, one will end up with a different 'picture' of the world, even with the same input.[49] That is, one will inevitably end up with different *models of data*. To compare the data to the theory, Giere (2006, pp. 68–69), following Suppes, argues that we only compare *models* of theory to *models* of data. That means, just as models of data represent a specific perspective of an object, so do models of theory represent a

47 Furthermore, many telescopes do not even measure wavelengths within the visible range. And all the more for it because they give us information on the universe that we could not otherwise gain.

48 The Compton Gamma Ray Observatory, which Giere gives as an example, was abandoned in 2000, but other observatories operate on similar principles.

49 This, of course, is not just the case for telescopes but all kinds of scientific instruments. Another example that Giere discusses are brain scans. Depending on which technology (CAT, PET, MRI, etc.) is applied, a distinct image of the brain depicting different aspects of the brain is obtained (Giere, 2006, pp. 49–57).

specific point of view as well. A model is never a complete replica of the object or phenomenon it represents. It only has a *fit* that we consider good enough given the circumstances. Depending on the use of a model, the requirements the model has to fulfil will change. Giere (2006, pp. 72–81) compares this to the way we use maps. The same place can be displayed in very different ways depending on the map's purpose. Take, for example, the task of mapping the earth. Transferring a spherical object onto a two-dimensional map will naturally cause a problem for the geographer. While the well-known *Mercator projection* serves the purpose to give sailors a navigational tool, it also drastically distorts the actual relative size of different countries and continents depending on where they are located on the map. Greenland, for instance, appears to be more than 14 times bigger than its actual relative size, making it as large as the whole continent of Africa. Alternatives like the *Peters projection* or the *Robinson projection* correct for this particular problem but on the flip side have to make concession with respect to other aspects. The *Peters projection* shows the landmasses at its right proportions but with distorted shapes. The *Robinson projection* tries to combine the advantages of both those maps as best as possible but does so by curving the longitude lines and would, therefore, not be very useful for navigation. Thus, each of these different maps is useful for different objectives. They all show a unique perspective of the world: "representation is representation *for a purpose*" (Giere, 2006, p. 80). None of these maps can give a fully accurate representation of all aspects of the surface of the earth. Maps are not copies of the place they display. But then again that is, arguably, not the point of maps. Maps just like scientific models are, as Giere points out, tools that represent the world in a certain respect, in a way that is helpful to our specific (scientific) endeavour. This is by no means giving in to total relativism, as Giere emphasises. Scientists can very well determine that one of two of the same kind of instruments is faulty if they produce completely contradictory data.[50] Equally, when an object can be detected from different perspectives (e.g., with different instruments), then this can be understood, argues Giere, to mean that there

50 To stay with Giere's example of the telescopes: if we were to point a gamma-ray telescope and a radio telescope at the same object and both deliver different observation, this would usually not be seen as a sign that one of the instruments is malfunctioning. But if instead there were two gamma-ray detectors that both are supposed to measure within the same range of wavelength registering something differently, scientists would, of course, conclude that at least one of the instruments is defective.

is "good evidence that there is *something* there, but this need not to be understood as knowledge in an 'absolute objectivist' sense" (Giere, 2006, pp. 57–58).[51] Meaning we cannot find the *one* 'objectively' true perspective of the object in question, in the sense of a 'view from nowhere'.

3.2.3 Observations in climate science

The climate system is a global system. Collecting and processing climate data, thus, is a global task. Having access to global data is crucial in gaining knowledge and understanding of the climate system.

This is not a new insight. Predicting the weather has been an age-old human endeavour. Success in such diverse aspects of public life from farming to warfare are dependent on knowledge of how the weather will develop. In order to do so, since the 19th century scientists have tried to establish an infrastructure that would enable them to collect data on a global scale. It is, therefore, not surprising that meteorology – and subsequently climate science – were big data science early on and the first to develop "systems for producing globalists information" (Edwards, 2010, p. 24).

As noted in the introduction to this book, one line of argument frequently used by climate-change sceptics is that the models must be false because they seem to disagree with the observational data. Climate scientists, however, counter that observational data just as the models are affected by uncertainties.

Just as much as the climate system is complex, the observational data retrieved from it is also complex. Complex systems produce complex data in more than one way: for one, in respect to the amount of data and, for another, in respect to the methods of acquisition, processing and evaluation. Traditionally, observational data has often been viewed as providing a form of context-independent confirmation of theories. The example of climate science will show in the following how this separation clearly cannot be upheld in actual scientific practice. Observations are neither as independent nor self-

51 Giere applies a very narrow and specific definition of *scientific objectivity* here. As we have seen in Chapter 2.3, this term has a rich history and even today a variety of interpretations. It might, therefore, be very well possible that Giere's perspectivism is 'objective' in a different sense. An interpretation of *scientific objectivity* which does not imply a 'view from nowhere', similar to, e.g., Longino's (1990) definition of objectivity as something that is achieved through diversity and a plurality of perspectives, would be much more compatible with Giere's perspectivism.

vindicating as climate-change sceptics often claim. On the contrary, Paul Edwards shows that in climate science data is "model-filtered" and models are "data-laden" (1999, p. 437). The lines between theory or model, respectively, and observation are (at least to a certain degree) blurry. The complexity of the climate system lays bare the interdependency between those two sides of science traditionally treated as distinct. Creating global climate data sets requires much more than just 'collecting' data. Or to quote Edwards: "if you want global data, you have to make it" (2010, p. 321).

In the following, I will discuss what constitutes climate data, how this data is collected and processed as well as the difficulties arising in the process. I will then examine what this means for the evaluation of climate models with observational data and our understanding of the relationship between models and observations in general.

3.2.3.1 Climate data

What actually constitutes climate data and could we not just collect all weather observations from the last few centuries and be done with it? After all, weather observations have been made for centuries now.

Unfortunately, it is not that simple. One way to approach the difficulties of creating climate data is to look at the specific historical differences in the requirements for data in weather forecasting and climate science. As Edwards (2010, p. 292) points out what traditionally distinguishes one from the other is their purpose – in a nutshell it is a matter of speed versus stability.

The purpose of *weather data* is to forecast the weather of the next few days. Reliable, easily accessible data which 'arrives' within the time limits of the forecasting cycle is required. Weather data, which is retrieved only after the new forecast has been made, is of little use for making forecasts. By contrast, the purpose of *climate data* is to create a statistically useful account of the climate over a longer period of time. For this one does not so much require data that is accessible within a specific timeframe but shows consistency in the way it is collected over a long period of time.[52]

To visualise this difference, one might look at what meteorologist and climate scientists usually focus on when they discuss temperature. While meteorologists making weather forecasts are in search of the absolute temperature value of a specific moment in time, e.g., the temperature in Bochum tomorrow,

52 For a good overview of the specific (historic) differences between weather and climate data, see Edwards (2010, pp. 294–295).

climate scientists, on the contrary, are traditionally interested in temperature anomalies, i.e., the deviation from the average temperature of a reference period.

3.2.3.1.1 Observations and uncertainties

To establish a comprehensive picture of the change in the earth's climate, a huge amount data drawn from a variety of sources is necessary. Besides data from hundreds of land-based observational stations, climate scientists also rely, for instance, on ships and buoys, to obtain observational data from the ocean and radiosondes deployed on weather balloons, airplanes and satellites to get information about the climate from different altitudes (Chen et al., 2021, pp. 174–177). For insight into the state of the climate before the beginning of systematic observations in the 19th century, scientists also make use of proxy data, such as tree rings or ice-cores (Chen et al., 2021, pp. 177–178). The processes of creating homogenous data sets out of the different types of data are intricate and epistemically challenging undertakings.

In the following, some of the difficulties that arise from the complexity and plurality of the observational climate data will be highlighted. The objective is to show why the ideal of raw observations providing clear-cut, context-independent and 'objective' evidence of the validity of a model or theory is so detrimental, specifically in the context of climate science. As we will see, observations in climate science are distinctly theory-laden so that it can be really misleading to even speak of such a thing as 'raw data'.

Coherent, long-term observations of the past climate are paramount to understanding how the climate might change due to increased anthropogenic forcing.

Although first attempts of infrastructural coordination of weather and climate observation date back to the middle of the 19th century, up until fairly recently, the need of weather data has often taken precedence over climate data, as Edwards points out (2010, p. 287). The value of consistent long-term climate data was only fully appreciated in the second half of the 20th century. Scientists, for the most part, just did not foresee that there would be the need for dependable long-term data in the future and observations records were often not kept. Traditionally, meteorologists seldom had use for 'old' data in the process of weather forecasting and for a long time storage space for that amount of data had been very expensive.

Figure 4: Change in percentage of possible sampled area for land records (top panel) and marine records (lower panel). Land data come from GHCNv3.2.0 and marine data from the ICOADS in situ record

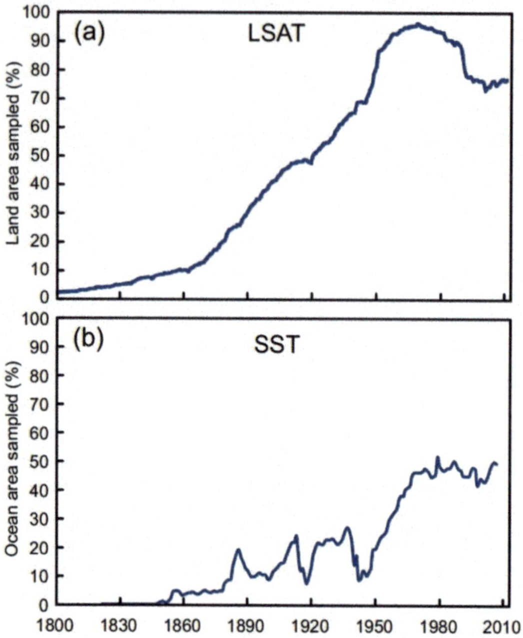

Source: Hartmann et al., 2013 p. 2SM-14 , Figure 2.SM.2

There are also often critical gaps in the recordings. Major historical world events can disrupt data records. There is, for instance, a lack of sea surface temperature (SST) data for the time of both world wars; see Figure 4. Most recently the COVID-19 pandemic has affected the amount of specific types of observational data collected during this time period because of a drop in air travel and ship traffic as well as interruptions in the regular maintenance of instruments. While the full effects of the pandemic on climate data records is not yet fully known, the consequences might not as bad as originally feared (Chen et al., 2021, p. 212).

Even when there is historical data at hand, there might also be gaps in information about the circumstances under which the data was collected, namely

which specific instruments were used and whether there were environmental and structural circumstances that might have systematically influenced the data in one way or another.[53]

Fortunately, many inconsistencies are random and, therefore, cancel one another out over time. When, for instance, the temperature is measured at a significantly large number of stations, measurement errors will most likely deviate in both directions. But sometimes inconsistencies have systematic causes. One example is the overall change of position of instruments in the Alps during the 19th century from on and at buildings to open space (Böhm et al., 2001; Edwards, 2010, p. 299). Another case are systematic changes to the instruments used. A prominent example for this is the change in material used for the buckets employed to measure ocean temperature. The 'evolution' of buckets, from simple wooden ones, over canvas buckets[54] to modern insulated ones meant that there were temperature differences up to 1 °C, depending on seasonal and local variables (Folland and Parker, 1995).[55]

The specific circumstances under which observations were made can be difficult to reconstruct. In certain cases, scientists can refer to metadata, such

---

53 For examples from sea surface temperature measurements, see Kennedy (2014).

54 The wooden buckets are actually "relatively well insulated and tend to have larger volumes leading to smaller temperature changes" compared to buckets made out of canvas (Kent et al., 2010, p. 719).

55 There are also other factors contributing to temperature differences in bucket measurements that have to be corrected for, such as "the size of the buckets (inner diameter and initial water depth for the canvas bucket model and bucket wall thickness for the wooden bucket model), the time the bucket was exposed on deck, the relative wind speed (which depends on the ship speed, the true wind speed and the degree of sheltering of the bucket) and the exposure of the bucket to solar radiation, all of which may vary from ship to ship and with time" (Kent et al., 2010, p. 723).
In some cases, measurements of SST are not taken by buckets but also through so called *engine room intake*, i.e., the seawater that is ultimately used to cool the engines of the ship. This kind of measurement also has a warm bias compared to buckets. Thompson et al. (2008) note that changes in the ships country of origin in 1945 led to an apparent temperature drop in SST: "Between January 1942 and August 1945, 80% of the observations are from ships of US origin and 5% are from ships of UK origin; between late 1945 and 1949 only 30% of the observations are of US origin and about 50% are of UK origin. [...] in August 1945 US ships relied mainly on engine room intake measurements whereas UK ships used primarily uninsulated bucket measurements" (Thompson et al., 2008, p. 648).

as user manuals of the instruments used and the likes, but those are not necessarily preserved (Edwards, 2010, pp. 317–319). Lack of this kind of information can constrict the accuracy of the 'picture' that can be drawn of the climate of the past from instrumental records.

Edwards (2010, pp. 17–28) argues that the creation of global-climate data sets is a question of infrastructure and insight into this infrastructure. Creating observational climate data sets requires having access to a globally organised network. From the 19th century onwards scientists slowly began creating a global observational network. Making use of this data requires climate scientists to do what Edwards calls an "infrastructural inversion" (2010, pp. 22–23): they have to turn the infrastructure 'upside down' to assess how the data was originally produced. However, even with an improved observational infrastructure, the problems described above are not purely issues of the past, as Edwards argues:

> Weather stations come and go. They move to new locations, or they move their instruments, or trees and buildings rise around them, or cities engulf their once rural environs. They get new instruments made by different manufacturers. Weather services change their observing hours and their ways of calculating monthly averages. These and dozens of other changes make today's data different not only from data collected 20 years ago, or even (sometimes) last week. It's like trying to make a movie out of still photographs shot by millions of different photographers using thousands of different cameras. Can we reconcile the differences, at least well enough to create a coherent image? Yes we can, scientists believe. But it isn't easy, and it is never finished. (Edwards, 2010, p. 6)

All in all, the need for long term, stable and global data make the creation of climate data sets a far form straightforward affair. Just collecting millions of single data points does not suffice; a great deal of data processing has to be done in terms of, reconstructing and homogenisation. Furthermore, climate modelling often requires gridded data meaning that the data points are spatially evenly distributed on a (virtual) grid. How this is done in practice calls for further methodological choices diversifying the approach that scientists can take (Parker, 2018).

3.2.3.1.2 Satellite data

Satellites as meteorological and climatological measurement facilities are a relatively recent invention. Images taken with the help of satellites were first used in the 1960s. But at first they were difficult to interpret and only marginally useful to local forecasting. Nevertheless, they were helpful in science communication as they provided visual aids for weather reports on TV, as Edwards (2010, p. 274) recounts. At the end of the 1960s, satellites carried the first instruments (radiometers) that were installed specifically to provide data for weather predictions. But it took a while until scientists had learnt how to implement this new data source into numerical weather predictions. Not until the 1990s, they actually substantially improved weather forecasting.[56]

These days satellite data are an essential part of weather forecasting and climate-change assessment. In the last decade, satellite data, notably a controversy about data from satellites equipped with so-called *microwave sounding units* (MSU), has also received some explicit attention from philosophy of science (see especially Edwards, 2010, pp. 273–279, 413–418; Lloyd, 2012). Even within the intricate sphere of climate data analysis, satellite data can be particularly complicated to 'read', which made integrating this data resource into meteorology and climate science such a difficult undertaking in the first place.

First of all, many instruments mounted on satellites only provide indirect measurements. Instead of temperature, MSU measure the microwave radiation emitted by oxygen molecules (radiance), from which then under the premise of a variety of physical and mathematical background assumptions the temperature of different layers of the atmosphere can be inferred.[57] Further, adjustments have to be made not just to filter out noise from the stratosphere but also to account for methodological issues. Especially considering that the MSUs are sequentially calibrated, the effects of orbital and instrumental decay have to be factored in (Wentz and Schabel, 2000). All of this comes with a variety of uncertainties and in practice requires complicated algorithms to account for those. That is, the 'raw' data is open to some degree

56 For a more detailed account of the history of satellite data in weather forecasting, see Edwards (2010, pp. 274–276).

57 To be more specific, the different frequencies of radiation are measured in distinctive 'channels' that can then be related, in a non-trivial way, to the temperature of different 'layers' of the atmosphere. In the case discussed here, scientists were interested in 'Channel 2' that measures radiance of the troposphere with some noise from the stratosphere.

of interpretations so that depending on the decisions made in this process different data sets can be obtained.

In the 1990s, an argument arose among climate scientists about the warming in the tropical troposphere. Models predicted that the tropical troposphere would show significant warming due to the increasing greenhouse gas emissions of the 20th century. But scientists Roy Spencer and John Christy (1990) claimed that the satellite data set developed by their research group, known as the UAH (*University of Alabama at Huntsville*) data set, evaluated with the help of radiosonde data actually disproved this. They argued that the radiosondes (on weather balloons) provided particularly reliable data as they are, contrary to satellite data, actually equipped with thermometers and measure the temperature of troposphere directly. However, other climate scientists, instead of discarding the model, questioned the reliability of the radiosonde data for the purpose of validating satellite data (e.g., Gaffen et al., 2000; Santer et al., 1999). They note that radiosondes are, in fact, prone to inconsistencies because they are exchanged frequently and their distribution is patchy. Lloyd (2012) points out that, contrary to what Christy and Spencer seemed to imply, radiosonde data does not provide a direct representation of 'reality'. The apparent evidence of cooler temperatures in the tropical troposphere that the radiosonde and UAH data sets were showing were misleading. Reconstructing MSU data in accordance with the radiosonde data set does not supply 'independent' evidence for the correct interpretation of the satellite data but relies instead on the (false) background assumption that radiosonde data could provide such.

In fact, it turned out that the same satellite data taking into account all methodological uncertainties could be processed in a way that created data sets that actually were much more compatible with the models: as was done in the case of the RSS (*Remote Sensing Systems*) and UMd (*University of Maryland*) data sets (Mears et al., 2003; Vinnikov and Grody, 2003). Eventually the dispute was settled, as far as the wider scientific community was concerned, at least to the extent that considering all uncertainties (observations and models) "there is no reasonable evidence of a fundamental disagreement between tropospheric temperature trends from models and observations" (Thorne et al., 2011, p. 66). However, research into these uncertainties continues to reduce the underlying

shortcomings of both models and data sets (Chen et al., 2021, p. 175; Santer et al., 2017).[58]

3.2.3.1.3 Paleoclimate data and proxies

For information about the climate before the beginning of structured instrumental recording, scientists make use of so-called proxy data. This term refers to "any biophysical property of materials formed during the past that is interpreted to represent some combination of climate-related variations back in time" (IPCC, 2021a, p. 2245). From the width of tree rings, for instance, it is possible to infer whether it has been a particular warm or rainy year. Similarly, scientist can gather information about the state of the climate many millennia ago from the amount of oxygen, and the distribution of dust particles or pollen trapped in ice through ice-core drilling. Further scientists also make use of historical documents that go beyond weather station records.

Diaries, farmers' and ship logs, travellers' accounts, official documents and newspaper articles may provide information not just directly on the weather but also information on times of harvest, crop yield, droughts, frosts or vegetation in general which can give indications of past climate developments (Chen et al., 2021, pp. 177–178).

Further, recently, indigenous knowledge has been recognised more and more as a source of information, such as, e.g., Aboriginal knowledge about sea level rise in Australia passed on through oral traditions over 7000 years (Nunn and Reid, 2016).

So there are a number of sources for paleoclimate data even when there are no direct observational records, presenting scientists with information about climate variables some going back millennia. Still, particularly proxy data must be interpreted with care and reconstructing the climate of the past with the

58 Santer et al. (2008), however, also concede that it might never be fully possible to solve the discrepancies in the observational data sets: "We may never completely reconcile the divergent observational estimates of temperature changes in the tropical troposphere. We lack the unimpeachable observational records necessary for this task. The large structural uncertainties in observations hamper our ability to determine how well models simulate the tropospheric temperature changes that actually occurred over the satellite era. A truly definitive answer to this question may be difficult to obtain" (p. 1719). Note, however, that this disagreement between models and observations is not considered by Santer et al. to mean that the models are necessarily wrong. The authors much more emphasise the uncertainties in observations.

help of proxy data hosts a variety of challenges. Here are just some of them (Frank et al., 2010; Parker, 2018; Schmidt, 2007):

For one, it can be difficult to accurately date proxy data. Some types of data has a yearly resolution (such as tree rings and ice cores), others can only be dated on a decadal scale (e.g., some pollen records or ocean sediment cores). Even with the more precisely datable data, there can be issues regarding allocating those data points to specific years, such as when it comes to the interpretation of tree ring growth (Mann, 2018).

Furthermore, most proxies are not equally well locally distributed. Ice-core data can only be sourced at the poles and tree ring growth is subject to seasonal differences.

There is also the issue that some proxies can be impacted by more than one factor. Plant growth for example, can be affected by temperature but also changes in soil and precipitation etc. So it is up to the scientists to figure out how to interpret the data and to find proxies that are less likely to be influenced by other factors.

And although proxy data is often calibrated against instrumental records of the recent past, sufficient instrumental records are only available from the last few centuries onwards, and the climatic circumstances of the earlier past might be outside the range of what we have instrumental records for.

While "sparse and noisy data are likely the underlying cause for the high methodological sensitivity" (Frank et al., 2010, p. 510) in paleoclimate records, proxies provide invaluable insight into past climate developments beyond instrumental records.

3.2.3.1.4 Reanalysis data

To fill the gaps of 'traditional' climate data, a new idea arose in the early 1980s to collect all available data of the last decades (or even centuries) and feed it into a weather model. The hope was to create a new, full-fledged data set that would provide a full "history of the weather, at every altitude, every grid point, every place on the planet" (Edwards, 2010, p. 323). After some years of searching for and assembling of data from all over the world, the first reanalysis projects started in the 1990s.[59] For this, climate scientists put this data retrospectively through a 4-D data assimilation as originally developed for weather

59 For a more detailed historical account of the development of the first reanalysis projects from the idea to execution, see Edwards (2010, pp. 323–336).

forecasting.[60] The models used for this have to be frozen so that changes and improvements that are consistently done in weather modelling would not interfere with the process. The first data sets coming out of reanalysis projects covered only a timespan of between five and 35 years (Edwards, 2010, p. 326). These days reanalysis projects like version 3 of *20CR* (short for *Twentieth Century Reanalysis*) of the *National Oceanic and Atmospheric Administration* (NOAA) and the *Cooperative Institute for Research in Environmental Sciences* (CIRES) produce data for the years between 1836 to 2015 (Slivinski et al., 2019).[61]

Using data that has been produced by heavily relying on weather models to evaluate climate models has provoked initially some concern of philosophers that one would run into a problem of circularity "since weather-forecasting

---

60 *Data assimilation models* originated in weather forecasting. The predecessor of these models, before numerical weather predictions (NWP) 'were a thing', was the so-called *analysis*. This consisted of handmade plotting of current data on maps from which scientists were literarily 'drawing', relying on their knowledge and expertise of the weather system, the forecast. With the introduction of NWPs gridded data became necessary. This gridded data was first produced by interpolating from the observational data by hand. Eventually, however, scientists began investing into so-called 'objective analysis', i.e., algorithmic process of interpolation. Then scientists started to integrate NWP forecasts as a 'first guess' for the time of observation. This had the advantage that scarcity of data in certain regions could be counterbalanced. Combining observations and forecast in *data assimilation* meant that uncertainties in model and data could be weighted and factored in. Soon scientists moved from three-dimensional assimilation to adding a fourth dimension: 'time'. This opened up the opportunity to integrate data lying outside of specific 'observing hours'.
Edwards emphasises that data assimilation has become much more than a "sophisticated version of interpolation" considering that "[a]ssimilation models are full-fledged atmospheric simulations; if run with no observational input at all, they would keep right on going day after day, month after month, generating physically consistent global data images. Where observations are available, they constrain the model, but they do not determine their output in any ordinary sense of 'determine'" (Edwards, 2010, p. 279). For a more detailed account of the development of data assimilation, see Edwards (2010, pp. 254–273).

61 What timespan different reanalysis projects comprise depends, in practice, on the specific objectives, the available computing power and on what kind of data are used. The ERA-interim atmospheric reanalysis from the *European Centre for Medium-Range Weather Forecasts* (ECMWF) only goes back to 1979 when satellite data became available for assimilation (Dee et al., 2011). On the other hand, the 20CR data set, covering the years 1836 to 2015, is created assimilating only surface pressure observation in order to avoid issues of inconsistencies with the availability of observational sources for that timespan (Slivinski et al., 2019).

models include a number of assumptions about the physics of the atmosphere that are similar, if not identical, to those included in state-of-the-art climate models [...][and thus] the fit between reanalysis data sets and simulations of past climate [...] will be artificially inflated" (Parker, 2011, p. 587). Leuschner, however, argues that precisely the close relationship between climate and weather models provides a good argument not to be particularly concerned about reanalysis data insofar as "if an assumption works reliably in a weather model this can well be considered as an indicator for the adequacy of the assumption" (Leuschner, 2015, p. 370). She further notes that the assimilation models used for reanalysis projects are, for one, well-tested in their function in weather predictions and, for another, that reanalysis data are usually applied in conjunction with conventional climate observations.

Reanalysis data sets are now widely in use in climate science and are often treated and referred to the same as 'normal' observational data. From a philosophy-of-science perspective, Parker argues that reanalysis actually constitutes a form of complex measuring procedure, albeit it is a "measurement practice that is still under development" (2017, p. 294). She adds this caveat not because of a worry that there is something innately wrong with using computer simulations to produce data but because reanalysis data to date "are not subjected to a rigorous process of calibration that provides well-motivated uncertainty estimates" (Parker, 2017, p. 300) as one would expect for 'ordinary' measuring instruments and observation making. Parker particularly points out that part of the measuring process in reanalysis involves some difficulties that are specific to the use of computer simulations, such as numerical discretisation that is not part of conventional measuring practices and needs to be accounted for in order for reanalyses to be full-fletched measurement practices in their own right.

3.2.3.2 Model-data interdependency

What has been shown so far is that creating global climate data involves much more than simply collecting data from a variety of instruments at a variety of locations. These data sets are "models of data" (Suppes, 1962) and a great deal of work goes into constructing them. Before the 'raw data' is of any use to climate scientists, they have to be extensively processed. That is, climate data sets are, as Edwards calls it, *model-filtered* insofar as they are created with the help of "what we might call 'intermediate models' [...][which] include models of instrumental behaviour, interpolation techniques [...], techniques for automatic

rejection of anomalous data points" and so forth (Edwards, 1999, p. 450). 'Raw' climate data on its own is patchy, inconsistent and sometimes conflicting. This is also a well-established insight in climate science: "[m]odel-filtered data can be trusted to the extent to which the models used to correct and extend the data have been independently tested and are confirmed" (Baumberger et al., 2017, p. 6). Or, to paraphrase Edwards: climate data has to be made (2010, p. 321).

However, Edwards also points out that the relationship between models and data is distinctly interdependent. Climate models are also "data-laden" (Edwards, 1999). Models are not just evaluated with the help of observations, they also contain in themselves a fair amount of observations. Specifically, the development of parametrisations requires scientists to consult observational data (Guillemot, 2010). For this reason parametrisations are also sometimes described as having a "semi-empirical" (Edwards, 1999, p. 449) character.

Tuning is another way in which models become data-laden. In the process of tuning models are calibrate with the help of observational data. There is some specific concern that this relationship might be questionable when the models are tuned to the same data they are later evaluated against. This practice, dubbed *double counting*, has sparked a discussion among climate scientists and philosophers. Scientists on the whole seem at least sceptical that this procedure could be considered adequate. Mauritsen et al. contend that evaluating quantities addressed in tuning are "of little value" (2012, p. 3) and the fifth IPCC Assessment Report (AR5) states that "quantities that are tuned cannot be used in model evaluation" (Flato et al., 2013, p. 749). A common strategy to avoid double counting in practice is *data splitting*: to use one half of the data set for tuning purposes and the other half for evaluation (Baumberger et al., 2017). Philosophers Katie Steele and Charlotte Werndl (2013), however, argue that from the point of view of Bayesian confirmation theory there is little difference between relying on the same data for tuning and evaluation and conventional methods of testing of hypotheses. Other philosophers and climate scientists (Frisch, 2015; Schmidt and Sherwood, 2015) have subsequently voiced criticism of this view. Frisch argues that, tuning has some confirmatory power but still concludes that "fit with data not used in tuning is a superior test of a model's performance" (2015, p. 174).

Edwards has called the connection between models and observation "symbiotic" (Edwards, 1999, p. 453). Contrary to the traditional ideal, there is neither a clear separation between data and model nor a clear-cut hierarchy where one

of the two takes precedence. It is "a mutually beneficial but also mutually dependent relationship" (Edwards, 1999, p. 453):

> The picture that I hope is emerging here is that all knowledge about climate change depends fundamentally on modeling. It's not that there is no such thing as an observation separate from modeling. It's that putting together a trustworthy and detailed data image of the global climate – getting enough observations, over a long time span – requires you to model the data, to make them global. It's not that climate simulations models are perfectly reliable, any more than weather forecast models always get it right. Instead, it's that simulations already include a lot of data in their parameters; they are precisely not pure theories, but theories already partially adjusted to the conditions we observe in the real world. That's model-data symbiosis. (Edwards, 2010, p. 352)

This symbiotic relationship can be understood in two ways, Parker (2020) points out. On the one hand, it can be a mere reference to the instance that in general a model is created with the help of a data set which was created independently from this type of model, which in turn is used to process another kind of data set. But there is no direct, two-way exchange between one specific data set and one specific model. On the other hand, occasionally there are, Parker notes, also cases where the relationship is more direct – to the extent that one data set is created with the help of a model, which is then evaluated with the help of this specific data set.[62]

---

62 One possible example for such a case, according to Parker, would be a model used to create synthetic data, which is then used to evaluate homogenising algorithm for finding, non-climate-change related inconsistencies in observational data. A direct symbiotic relationship may arise when the model producing the synthetic data is later evaluated against an observational data set, which was created with the help of the homogenising algorithm, which in turn was tested with the synthetic data. This kind of symbiotic relationship in general, argues Parker, does run a particular high risk of turning circularly in a self-affirming way, as the model "has no direct role in producing the climate data set; it merely plays a supporting role in efforts to evaluate methods for removing artifacts when producing the data set" (Parker, 2020, p. 815). Parker comes to the conclusion that reanalysis is one case where one might have more reason to be concerned that this relationship is problematic, to the extent that as "weather-forecasting models (used in data assimilation) and climate models take a similar approach to representing physical processes in the atmosphere, it could be that reanalysis data sets and climate simulations have some shared errors" (Parker, 2020, p. 816). However,

3.2.3.3 Verification and validation

One issue that arises in the context of evaluation of climate model concerns the applicability of the concept of *validation* and *verification*. Traditionally, the process of *verification and validation* (commonly shortened to V&V) is drawn upon to establish confidence in computer simulations:

> *Verification* is said to be the process of determining whether the output of the simulation approximates the true solutions to the differential equations of the original model. *Validation*, on the other hand, is said to be the process of determining whether the chosen model is a good enough representation of the real-world system for the purpose of the simulation. (Winsberg, 2019)

Both are commonly treated as distinct in execution and in conceptual classification. While verification is considered to be a question of mathematics and the accuracy of the numerical solution, validation concerns physics and the question whether the underlying equations of the model are an adequate representation of the target system. Both pose two separate questions:

> First, are the solutions that the computer provides close enough to the actual (but unavailable) solutions to be useful? [...] Second, do the computational models that are the basis of the simulations represent the target system correctly? (Frigg and Reiss, 2009, p. 602)

Although this conception is popular with scientists (Winsberg, 2018, pp. 156–157), philosophers have raised concerns that applying these terms to scientific models is problematic because "it is impossible to demonstrate the truth of any proposition, except in a closed system" (Oreskes et al., 1994, p. 641): a requirement that only purely logical or mathematical models can meet.[63] There is some disagreement within the philosophy of science community about the extent to which this concern about the entanglement of verification

---

Parker notes that current research does not show that reanalysis data sets are 'closer' to models than conventional observational data.

63 Oreskes et al. come to the conclusion that in this instance instead of verification or validation the best that might be accomplished is *confirmation*, that is, there is an increasingly good match between increasingly diverse observations and the model output: "The greater the number and diversity of confirming observations, the more probable it is that the conceptualization embodied in the model is not flawed. But confirming observations do not demonstrate the veracity of a model or hypothesis, they only support its probability" (Oreskes et al., 1994, p. 643).

and validation holds in actual scientific practice and for which kinds of computer models (Jebeile and Ardourel, 2019; Morrison, 2015). Although it has been noted that practices concerning either primarily verification or validation are part and parcel of climate modelling (Winsberg, 2018, p. 157; Lenhard, 2018). Philosophers concerned with the peculiarities of complex computer simulations of the type of ESM have pointed out some features of these models that should make us sceptical about V&V as an epistemological concept of two separable procedures fully grounding our trust in these simulations. Why is that?

Lenhard (2018) argues that the applicability of V&V is limited in simulations of the type of global climate models because it would require to separate model structure and parameter. However, climate models necessarily also include parametrisations schemes with adjustable parameters, so that the adequacy of a model cannot be assessed without already having determined the parameter value. That is, "without assignment of parameters neither the question about representational adequacy nor the question about behavioral fit can be addressed" (Lenhard, 2018, p. 842). For this reason, he concludes that "[i]t is not possible to first verify that a simulation model is 'right' before tackling the 'external' question whether it is the right model" (Lenhard, 2018, p. 842).

Winsberg (2018, pp. 156–160) comes to the same conclusion from a slightly different angle. He points out that these models rather have a "life cycle" than undergo a "linear development" (2018, p. 158). Tuning, the need for parametrisations and a fuzzy modularity, means that the process of model development is an "iterative process" (Guillemot, 2010, p. 249; Winsberg, 2018, p. 158), where the model is consistently tested and further developed. New elements are added to the model, for instance, in form of new parametrisations and/or the discretization scheme is modified. That is, there are constantly changes made to both the underlying model and the implementation in a trial-and-error fashion, not just based on basic well-accepted physical and mathematical principles and theories, but also "physical intuition, phenomenology, local empirical finding, lore accumulated from parallel modelling successes, etc." (Winsberg, 2018, p. 158). Winsberg particular points out that this process leads to the possibility of unknown compensating effects so that:

> [w]hen a climate model succeeds at passing whatever test we subject it to, it might be because the underlying model is ideal and the algorithm in question finds solutions to that underlying model. Or it might be because of a "balance of approximations." This is likely the case when a model is delib-

> erately tailored to counterbalance what are known to be limitations in the schemas used to transform the model into an algorithm. [...] And when success is achieved in virtue of this kind of back-and-forth, trial and error piecemeal adjustment, it is hard to even know what it means to say a model is separately verified and validated. (Winsberg, 2018, pp. 159–160)

But Winsberg (2018, p. 160) also argues that this amalgamation of improvements efforts, constricting the model from different perspectives and objectives, are exactly what can substantiate the confidence that the models are adequate for a specific purpose.

Further, the notion that verification and validation cannot be kept fully separate does not mean that procedures targeting one or the other do not have an important place in the practice of climate modelling. As Lenhard notes the holism underlying the problem with V&V in complex computer simulations "comes in degrees" (2018, p. 842).

3.2.4 Conclusion

Contrary to conventional ideals about how science should operate this chapter has shown that in climate science the relationship between models and data is rather complicated and interdependent. For one, observational or experimental data is not just consulted at the end of the 'scientific process' to confirm or refute a theory or a model. Instead observations are a vital and intricate part in the climate-model building process. Not only are parametrisations often significantly based on observational data, but models are also continually calibrated, evaluated and further developed with the help of observations. These models are not just simply a representation of theory. They do not fit into the traditional theory-focussed narrative of science, whereby observations only play a minor character in the scientific process by verifying theories in the end.

Some philosophers like Hacking worrying that the experimental part of science was neglected have argued for the independence of the experimenter's work from that of the theorist's. However, when science is dealing with systems that are as complex as the climate system, it becomes increasingly undeniable that this separation into an empirical and theoretical part of science can no longer be maintained. Climate models are not just pure theory. They are laden with data through parametrisation schemes, tuning and evaluation processes. Similarly, observational data are clearly theory-laden. The pure, 'raw' measure-

ment, the direct instrument reading without any processing, usually says little about the quality of the model.[64] Theory and observation are neither fully independent nor does one take precedence over the other. As Edwards (1999) has rightly pointed out, models and data have a "symbiotic" relationship. Importantly, this should not be confused with the claim that models and data would be interchangeable or in some way 'the same': "Interdependence is not identity; data sets are still derived *primarily* from observation, and models *primarily* from theory" (1999, p. 454).

Nevertheless, when the assumption of 'independent observations' providing irrefutable evidence for a theory or model is widespread in the public understanding of science, it makes it easier for science sceptics or particular interest groups to undermine trust in scientific research. Edwards (2010) and Lloyd (2012) have shown that in the case of the controversy about the interpretation of the MSU satellite data discussed above, the debate was not just held within the scientific community but also eagerly picked up by climate science sceptics from politics and the wider society as an argument against the reliability of climate models. The claim that the UHA data set, because it was calibrated with the help of radiosonde data, could serve as a kind of independent representation of 'reality', functioning as a benchmark against which the quality of models could irrefutably be judged, was also brought forward in hearings in front of the House Representatives in the United States. And the apparent 'mismatch' between the models and observations was presented as a failure of the scientists to do "sound science" (Edwards, 2010, p. 414, see also Chapter 3.3.1).

Instead what the case of the MSU data has shown is that, taking into account that observations just as models come with some degree of uncertainties, one mismatch with some specific data set does not necessarily mean that models ought to be disregarded right away.

Considering the controversy about the UHA data sets Lloyd (2012) introduces the distinction between *direct* and *complex* empiricism. From the point of view of direct empiricist 'raw' data are "windows on the world, as reflections

64 In fact, Edwards notes that the meaning of *data* itself has changed in climate science. In the early days of climate modelling, scientists used to separate between data gained from simulations and that coming directly from observations. But this turned out to be "linguistically awkward" (Edwards, 2010, p. 283), so it became common practice to refer also to model output just as *data*. While many philosophers have been critical about the term *raw data* in general (Harris, 2003; Leonelli, 2019), computer simulations and techniques like reanalysis have expanded the definition of 'measurement' (Parker, 2017).

of reality, without any art, theory, or construction interfering with that reflection. This claim of a direct connection to reality is very important to their views" (Lloyd, 2012, p. 392). Lloyd argues that instead a "complex empiricism", which allows for the idea that data is theory-laden, in need of extensive processing and is to a certain degree open to interpretation, would be more appropriate.

But, of course, despite all these difficulties, climate models are, nonetheless, broadly and continuously evaluated in respect to their fit with observations. But as observational data sets are, on the one hand, model-filtered and, on the other hand, models are data-laden in terms of tuning and semi-empirical parametrisations, it can be difficult to determine what a 'good' fit of models with observational data actually means. In this respect, climate scientists set much store by the *adequacy-for-purpose* principle (Parker, 2009; see also: Notz, 2015; Knutti, 2018; Chen et al., 2021, p. 221). While in general the models constantly get better at representing the climate of the past, they, nevertheless, do not, as we have seen in Chapter 3.1.3, display all relevant variables and processes equally well. So it has to be established what specific variables and processes are relevant for the model to be adequate for a specific purpose. Thus, "[t]he challenge [...] is to determine which instances of fit do support and which instances of misfit do undermine an adequacy-for-purpose hypothesis" (Baumberger et al., 2017, p. 6). However, even if the properties of fit a model is required to display for a specific purpose are determined, a model showing a good fit in this respect still does not necessarily warrant that the model will be adequate for making predictions about the future, as:

> [i]nstances of fit could be the result of compensating biases, or overfitting, or could simply be unimportant if the evaluated quantity is unrelated to the prediction of interest. Instances of misfit could result from the fact that the model simulates a different quantity than that observed or from biases in observations, or of different processes in models and observations. (Knutti, 2018, p. 331)

This means, that it is not possible to simply extrapolate from an adequate representation of current and past climate that the models will necessarily be able to represent future climate states with high anthropogenic forcing equally well. That is, the climate of the future might lie outside the boundaries

of what we currently have data for, so that feedbacks[65] or other processes might emerge that are not accounted for by evaluating the model with regard to their fit to the available data (Baumberger et al., 2017, pp. 8–9).[66] Further, it might very well be that (unknown) compensating effects within the model contribute to a good fit to observations that might not hold under future climate change conditions because, for instance, feedback processes might accelerate in unforeseen ways.

On the other hand, when a model is not able to represent a feature of the climate of the recent past, then it can be said that it is highly unlikely that the model is adequate for projecting that feature of the climate in the future (Parker, 2009). Therefore, a good fit of the model with observational data has become a necessary but not sufficient condition for the model to be adequate for the particular purpose. In other words, "empirical accuracy of model results should […] be understood as premises" (Baumberger et al., 2017, p. 7).

For these reasons, a good understanding of different climate processes and the system as a whole are considered by climate scientists key to estimating to what extent these processes are adequately represented in the models (Bony et al., 2013, p. 20). Based on this assumption, scientists specifically highlight the necessity of a proper understanding of the inner workings of the models themselves when one wants to draw any conclusions about the models' applicability for climate change predictions:

> We need to make sure the models do the right thing for the right reason, because we want to use them beyond the range they have been evaluated. We have greatest confidence in models where we understand the processes behind the results, and where we can argue that models represent them well enough. (Knutti, 2018, p. 346)

Knutti also argues that "process understanding" (2018, pp. 334–338) is central to the adequate-for-purpose question. That means having an insight into whether the emergent component of the model arising out of the inner model

---

65 Climate feedbacks refers to those processes in the climate system where a change in quantity *a* also impacts another quantity *b*, which leads to further change in quantity *a*. In general these interactive relationships can, depending on the specific process, have an accelerating (positive feedback) or decelerating (negative feedback) in the context of climate change. Important feedbacks concern, for example, clouds, the carbon cycle and ice-albedo.

66 One way that climate scientists deal with these problems is establishing so called *emergent constraints* (this will be further explored in Chapter 3.3.3.4).

structure, are similar to the target system. Further, process understanding requires that the interactions of the emergent elements in the model are well enough understood, so it is reasonable to believe that, firstly, it will remain like that over a significantly long period of time and outside the range that can be evaluated and, secondly, that no other relevant feature is missing. Only if all of this is accomplished, argues Knutti, one could consider a model adequate for purposes that lie outside the range for which the models can be tested. Thus, mitigating the epistemic opacity of the models through background information that climate scientists have acquired both about the inner-workings of the model and the target system and understanding to what extent specific climate processes are adequately represented in the model is seen as an essential aspect in improving climate model projections (see also Baumberger et al., 2017).

All of this requires much more than pure model-data comparison. Instead, as will be further examined in Chapter 3.3.3.4, what makes climate scientists confident in using models for specific purposes are usually a variety of different factors that go well beyond simply being in accordance with observations; the model fit is just one aspect among many.

To summarise it can be said that what has been stated in this chapter about the relationship between the empirical and theoretical part in science holds, specifically for instance, where science deals with highly complex systems. But it would be an illusion to say that these two spheres of science can be fully separated, even when science is concerned with less complex systems. And as science turns to more and more complex problems, this will just become more and more obvious. On the other hand, as we have seen Chapter 3.2.2, philosophers and sociologists have long been pointing out the complex role of observations in science and climate science is here no different than other sciences (Guillemot, 2010).[67]

One reason why the ideal that observations provide a readily available irrevocable benchmark against which theories can very easily be evaluated has to be assumed to be rooted in the fact that so far the need for and difficulties of processing and handling of data has not been particularly noticeable from

67 Guillemot particularly notes that compared to many other sciences the "data malleability is of a much higher degree in the climate sciences, due to the extensive use of computers on all levels" (Guillemot, 2010, p. 249).

outside of science. The wish to clearly separate one from the other is understandable. Vindicating climate models would be much easier and debates about the specific impact as well as scale of anthropogenic climate change could be settled much quicker if there was some kind of irrevocable method of providing evidence in the form of observations telling us instantly if the models were right or wrong. If we come to the table with these expectations, then it seems altogether unsatisfying to hear that the only possible statement about the model is whether increased confidence in the adequacy of the model for a specific purpose is warranted. But things are rarely as neat in science as outsiders and often even scientists themselves wish them to be. Science is messy and complicated. This is not a denial of climate change nor a dismissal of science itself. As illustrated above in the case of the MSU data, even in situations where there is disagreement on the interpretation of the data at first, over time a consensus will be found within the community. Scientists, in the end, are often able to settle these kinds of debates. But this takes time as well as effort and will always be accompanied by some (never fully disappearing) uncertainties.

3.3 Predictability

3.3.1 Introduction: predictability and uncertainty

The wish to predict the future, to produce forecasts of what is to come is an age-old human endeavour. In ancient Rome, animals, specifically birds, were observed to divine some knowledge about the will of the gods. In ancient Greek, the oracle of Delphi was consulted, before wars were fought, and astrological readings of the stars can be found in many cultures going back millennia.

As far as science is concerned, making predictions as an integral, central function of science turned into an essential part of the definition of science in the 17th century, including "the conviction that any discipline that does not make successful predictions thereby fails to make good its claims to providing scientifically adequate understanding" (Rescher, 1988, p. 25). The emphasis on prediction making in this conception of science only gained in intensity over the next centuries with Newtonian physics and enlightenment being driving forces. The hope was that science would progress to supply ever more precise predictions. However as Rescher notes, the rise of the predictability ideal in science, took a hit in the wake of the French Revolution and only came back "as

science of nature gave way as source of inspiration to the science of society" (Rescher, 1988, p. 25) in the 19th century.

In the post–World War II period, the public and political 'appetite' for predictions reached a new high (Rescher, 1988, p. 28). At the same time advancements in computer power and the rise of computer simulations in the second half of the 20th century increased the abilities of science to fulfil these expectations. The first weather models came out of the effort put into simulating the explosion of nuclear weapons at the beginning of the Cold War when John von Neumann recognised that the insight acquired there could also be applied to weather forecasting (Weart, 2010, p. 209).[68] The use of computer simulation spread to multiple other fields of science over the rest of the century. Computer simulations now have a wide variety of applications, such as economics, epidemiology, engineering, cosmology and much more. Further, the progress in computer modelling did not only meet the request for information about the future, it also facilitated it, as Heymann et al. point out: "models helped both to create and furnish social demands for predictive knowledge" (2017b, p. 6).

However, while at once the demand for predictions increased from the middle of the last century onwards, at the same time, with this increasing relevance of science in public life, the pressure for science to provide clear, irrefutable or, to be more specific, uncertainty-free predictions also increased. From the middle of the last century onwards it became a popular and very effective strategy of some interest groups to undermine research results inconvenient to them, by arguing that there are still too many uncertainties to take actions, and that, before any action can be taken, there first needs to be more research done (Oreskes and Conway, 2010).

A helpful way of thinking about these two somewhat contradictory expectations about what science ought to deliver is in terms of the concept of *cultures of predictions* as applied by Heymann et al. (2017a).[69] This framing will be use-

68 It still took a long time from that initial idea to any kind of weather simulation that could be used for actual forecasting purposes. Two names that should be mentioned here are Jules Charney, who was engaged by Neumann to lead the group that would develop the first weather model and Lewis Fry Richardson, who developed the numerical system based on Bjerknes equations (Chapter 2.1), which was used to build the first weather model (Richardson, 1922). For a longer recount of the history of weather and climate modelling see Weart (2010) and Edwards (2010).

69 The concept of "culture[s] of prediction" was not invented by Heymann et al. (2017a) as the authors themselves state. For instance, Fine (2010) first applies the phrase "cul-

ful to understand the way the social and the scientific spheres are interconnected through reliance on the predictiveness of science, the difficulties that come with the interdependency of these two spheres and how this specifically translates to the case of public perception and expectation of climate science. Heymann et al. characterise the cultures of prediction in five steps in order to show "the broad-ranging and pervasive role of predictive efforts in postwar modern society" (2017b, p. 6):

1. the social role of prediction;
2. the character and significance of computational practices;
3. the domestication of uncertainty;
4. the degree of institutionalization and professionalization of predictive expertise;
5. the cultural impact of predictive practices and claims (Heymann et al., 2017b, pp. 6–7).

Firstly, predictions fulfil a significant social role. The general expectations for science to provide knowledge about what the future will bring extend to most aspects of society. Heymann et al. (2017a, pp. 20–22) note that this also means that prediction making can sometimes take precedence over understanding the system, a prioritisation that is not uncontroversial within the scientific community. This is also a discussion that has been taking place in climate science (Bony et al., 2013). However, it has to be said that, as will be further discussed below, it is questionable whether understanding the systems and the simulations, and improvements in predictions making skills can be fully separated. Consequently, some climate scientists advocate to put the focus of climate science towards finding ways to mitigate the epistemic opacity of the models (e.g., Baumberger et al., 2017).

Secondly, with the vast demand for predictions, scientists increasingly rely on computer simulations (Heymann et al., 2017a, pp. 22–26). This comes along with specific epistemic challenges, many of which have already been discussed in the previous chapters (see particularly Chapter 2.1).

---

ture of prediction" in his in-depth study of American Weather Forecasting Institutions, while Johnson (2017) uses the term to explore the role of mathematics in prediction making. Heymann et al. say they specifically use the plural here to emphasise "the local origin and socially contingent character of the cultural formations built around the construction and use of computer models for predictive purposes" and that there is "a multitude of distinct cultures of predictions" (Heymann et al., 2017b, p. 6).

Thirdly, prediction making usually comes hand in hand with uncertainties science has to deal with alongside the public's cravings for certainty (Heymann et al., 2017a, pp. 26–29). The specific kinds of uncertainties climate modellers face will be discussed further in the following. As already noted in the introduction to this book, in the context of climate science uncertainties often get misrepresented in public debates as 'evidence' that nothing can be said about anthropogenic climate change at all. The question of how to deal with uncertainties also has to do with the first characteristic of *cultures of predictions* as climate scientists argue that uncertainties can be reduced by increasing background knowledge about the models and the target system (Bony et al., 2013; Knutti, 2018).

Fourthly, roughly in the last seventy years, the increasing demand for scientific predictions has also led to the creation of a variety of institutions whose primary concern is to provide predictive knowledge. In the case of climate science, arguably the most significant one of those institutions is the IPCC, whose role in assessing and communicating climate change will also be discussed below (Heymann et al., 2017a, pp. 29–32).

Last but not least, the focus on predictions in the last century also had a wider cultural impact. One prominent example is how much the public mindset has been directed towards climate change. The cultural impact and the power are such, argue Heymann et al. (2017a, pp. 32–36), that we are usually not aware how the cultures of prediction permeate everyday life and direct our view of the world.

Heymann et al. argue that "cultures of prediction represent cultures of power" (2017b, p. 7). However, they also note that the power attributed to science in its abilities to make predictions, does not necessarily translate to political actions. On the contrary, it can halter or even undermine political actions. In the context of climate science, this can be exemplified with help of the phrase 'sound science'. It relates to an argument against taking up a stronger climate-change mitigation policy, brought forward by climate science sceptics. That is that there are still too many uncertainties and that, unless they are eliminated, it would be an overreaction to act. As Oreskes and Conway (2010, pp. 136–163) have shown, the method to call for 'sound science' as a premise before any political action should be taken first arose as a strategy of tobacco companies in order to discredit research showing connections between second-hand smoking and cancer, and to avoid further regulations in the early 1990s. A strategy that was quickly adopted and reapplied to argue against, among other issues, regulations of CO_2 emissions, as Oreskes and Conway

further show (pp. 169–215). By the mid 1990s, the claim that climate scientists were not doing "sound science" but "junk science" found its way to hearings at the US House of Representatives and Senate (Edwards, 2010, pp. 411–414). In Chapter 3.2. we have seen that the demand for 'sound science', alluding that there is an apparent discrepancy between observations and models, can function as an embodiment of the ideals that observations provide irrefutable evidence for or against the models. On the other hand, the call for 'sound science' also puts the emphasis on the uncertainties that come along with making predictions. As Supran & Oreskes (2017) have shown in an analysis of internal communication of Exxon Mobile Cooperation insinuating that there is not yet sufficient evidence of global warming was a well-established tactic to thwart stricter climate policy. They conclude that in order to avoid regulations the company relied on a publicity strategy that "overwhelmingly emphasized only the uncertainties, promoting a narrative inconsistent with the views of most climate scientists, including ExxonMobil's own" in order to "undermin[e] public understanding of scientific knowledge" (Supran and Oreskes, 2017, p. 15).

Thus, while the post–World War II period increased the relevance of science in the public sphere and the expectation but also ability (through rising computing power) to provide predictive knowledge, it also brought with it the expectation of science to deliver binary, clear-cut and uncertainty-free predictions. Thereby, science risks becoming trapped in the conflict between the self-awareness that knowledge is preliminary and the public expectation to provide clear-cut answers to questions of the future. When the misconception that uncertainties are a sign of bad science or not fully matured science is widespread in the public perception of science, this can quickly become a problem. While scientists whose work is under public scrutiny, like climate science, are often very much aware that the complexity and the impact of their research does warrant caution, as the following will show, they also have to navigate, on the one hand, public expectations to provide meaningful answers and, on the other hand, the knowledge that any misplaced overconfidence on their part can be misconstrued as a sign of the overall corruption of their field of research.

Considering the structure of this subchapter, I will first discuss *robustness analysis* (RA) as it has become central to the debate in philosophy of science on how to deal with the uncertainties occurring in modelling and specifically in climate modelling. I will, then, go on to discuss uncertainties in climate science more explicitly, how they are communicated to policymakers and stakeholders and

how the argument about RA plays out here. This analysis of uncertainties and RA, on the one hand, will show that the sources of uncertainties climate scientists are confronted with are multifaceted and intricate so that the demand for climate science to reduce these uncertainties is often easier said than done. But, on the other hand, this does not mean that these uncertainties fully debilitates climate science and render it impossible to make any kind of statement about the way the climate will alter due to anthropogenic forcing. It will be shown how, despite all kinds of uncertainties, it is possible for climate scientists to assess specific hypotheses. This will be done by way of example of Equilibrium Climate Sensitivity (ECS) which is one of the most important variables to determine the effects of anthropogenic climate change.

3.3.2 Robustness

The notion of RA of models was first introduced in philosophy of science by Richard Levins (1966). As a biologist, Levins observes that many models in population biology, his field of research, include idealisations and simplifications due to the complex nature of the systems in question. This, in turn, means that the models also include elements that are not 'truthful' representations of the world.

Because the systems are often too complex to model these systems with a "naïve, brute force approach [...] which is a faithful, one-to-one reflection of this complexity" (Levins, 1966, p. 421) to be feasible in practice, the question arises how we can still infer knowledge about a specific phenomenon from the models despite these idealisations. That is, how do we know that a model result is due to "the essential of a model or [...] the details of the simplifying assumptions" (Levins, 1966, p. 423).

Levins' proposal for a solution to this conundrum is to use a variety of different models with different idealisations that, nevertheless, share a common core regarding the phenomena:

> we attempt to treat the same problem with several alternative models each with different simplifications but with a common [...] assumption. Then, if these models, despite their different assumptions, lead to similar results we have what we can call a robust theorem which is relatively free of the details of the model. Hence our truth is the intersection of independent lies. (Levins, 1966, p. 423)

Levins concludes that, when models complement each other to the extent that they are "coordinate alternative models for the same set of phenomena" (1966, p. 431) and generate matching results, this constitutes a *robust theorem*. Meaning a common, reliable prediction, even though every single model is wrong in respect to the representation of some aspect of the target system.

Levins' account of robustness has subsequently been criticised by Orzack and Sober for asserting "that a statement's *robustness*, as distinct from its *observational confirmation*, can be evidence for its truth" (1993, p. 538). They criticise Levins' concept of RA for seeming to provide, as Weisberg put it "a novel, nonempirical form of confirmation" (2006, p. 732). Orzack and Sober provide a more formalised account of RA. They begin by identifying the specific circumstances under which, they say, a clear relationship between robustness and truth can be established:

> There is a special case in which the connection between robustness and truth is clear. Suppose we know that one of a set of Models M_1, M_2, ..., M_n is true, but we do not know which. If *R* is a robust theorem with respect to this set, then *R* must be true. (Orzack and Sober, 1993, p. 538)

However, while in this specific situation the premise is that one of the models in the set is true, Levins' assumption is that all models are "lies" one way or another. Orzack and Sober argue, that unless we know that at least one of the models in the set is true, we cannot infer from the fact that the models predict the same that it is true.

Otherwise it might very well be the case that a robust theorem arises because of a common denominator in the form of a shared assumption within the model, which might or might not be a lie that all models have in common. From this point of view, a robust theorem might reflect more about the conveniences of the model building process than its truth (Orzack and Sober, 1993, p. 538).[70]

70 Orzack and Sober also consider what independence of models means. Like Levins they regard the independence of models in a set as a necessary premise for robustness (1993, pp. 539–540). They argue that there are two ways in which the models could be considered independent: logical or statistical. With regards to the first case, it has to be said that "competing models are not logically independent" (Orzack and Sober, 1993, p. 539), and with regard to the second case, there remains the question of how one would sample from the whole set of models. That is, reason Orzack and Sober, both kinds of independence cannot be applied to RA of models in scientific practice.

Levins, subsequently, has defended his approach to robustness from Orzack and Sober's attack by pointing out that RA as he understands it does not forego empirical observations:

> Orzack and Sober are worried that the robustness strategy seems to propose a way to truth independent of observation. This is not the case. Observation enters first in the choice of the core model and the selection of plausible variable parts, and later in the testing of the predictions that follow from the core model. (Levins, 1993, p. 554)

The way RA is represented by Orzack and Sober does not take this into account, Levins argues.[71] He notes that seeking robust theorems "reflects the strategy of determining how much we can get away with not knowing, and still understand the system" (1993, p. 554), though as Weisberg points out Levins "does not tell us how it helps to confirm models and their predictions" (2006, p. 732).

Building on this discussion, Michael Weisberg (2006) offers a new approach to RA. Weisberg's goal is to show that "robustness analysis is effective at identifying robust theorems, and while it is not itself a confirmation procedure, robust theorems are likely to be true" (2006, p. 732). To do so requires a more differentiated understanding of the concepts of *robustness theorem* and *robustness analysis*, argues Weisberg.[72] To that end he proposes to see RA as the following four-step procedure (Weisberg, 2006, pp. 737–738):

1. Finding a robust property that is a shared result among an ensemble of models
2. Studying the models to find a common structure that creates the robust property

---

71 Lloyd, whose account of RA in the context of climate models will be discussed further below in Chapter 3.3.3.4, argues that the differences between Orzack and Sober, and Levin are the divergent objectives: "Orzack and Sober had a different goal, namely predictive inference to the model's outcome [...] about which they were likely correct [...]. Levin, in contrast, emphasised the key empirical evidence for the model structure under consideration" (Lloyd, 2015, p. 59).

72 Weisberg presents his concept of RA in the context of the example of a predator-prey model. I will just introduce his argument here in an abstract form as I will discuss an application of Weisberg's four-step robustness scheme to climate modelling by Lloyd (2015, 2010, 2009) in Chapter 3.3.3.4.

3. Determining how the given mathematical model structure is to be empirically interpreted
4. Undertaking stability analyses by examining how the robustness theorem will fare if the models in the ensemble change somewhat

The first two steps, argues Weisberg, go hand in hand. Once (or often while) a robust property (that is a common predictive result the models generate) is found in "a sufficiently diverse set of models" (Weisberg, 2006, p. 737), the core structure which brings about the robust property has be determined. Weisberg calls this the *common structure*. In the most basic case, the common structure has the same mathematical form in every model, but this is not necessarily the case.[73] After the first two steps a specific mathematical description has been obtained but not yet any connection to an empirically observable phenomenon has been made. This follows in the third step. Without determining this empirical description, the robust property might as well be just a mathematical construct to be found in all models but does not tell us anything about the real-world system we are interested in, Weisberg points out.

These three steps culminate, he argues, in the formulation of the *robustness theorem*, which has the following general conditional form:

> Ceteris paribus, if [common causal structure] obtains, then [robust property] will obtain. (Weisberg, 2006, p. 738)

In a concluding fourth step of Weisberg's definition of RA, different kinds of stability analyses are performed in order to investigate what happens to the robust theorem when the circumstances characterized in the models change slightly.

This is, according to Weisberg, the four-step process of RA. But Weisberg also states that RA has some epistemic power. However, he emphasises that, while RA is applied in science to further knowledge and understanding about real-world phenomena, determining a robust theorem by itself is not sufficient:

---

73 Weisberg adds that, in those cases where the common structure has not the same mathematical form, it can make the identification and analysis of the common structure much more difficult and may lead to situations "in which theorists rely on judgment and experience, not mathematics or simulation, to make such determinations" (2006, p. 738).

> A common reason theorists engage in robustness analysis is to increase the quality of their predictions and explanations about real phenomena. Although useful for both of these purposes, the theorems generated by robustness analysis cannot fulfill either role alone because they are conditional statements, further attenuated with ceteris paribus clauses. Explaining a real-world phenomenon or predicting its occurrence requires us to know that the common structure is actually being instantiated and that no other causal factor is *preempting* the efficacy of the common structure. (Weisberg, 2006, p. 739)

To do so, one would usually turn to empirical testing. However, Weisberg notes that RA is often done in instances where options to do so are lacking. He sees RA as a procedure that under certain circumstances is still informative and "can give us good reasons to believe the predictions and explanations of robust theorems" (Weisberg, 2006, p. 739). To be, thus, epistemically informative, two questions have to be answered, Weisberg argues:

1. How frequently is the common structure instantiated in the relevant kind of system?
2. How equal do things have to be in order for the core structure to give rise to the robust property? (Weisberg, 2006, p. 739)

In the absence of empirical data, the first question can be answered, at least to a certain degree, by making sure that a "sufficiently heterogeneous set of situations is covered in the set of models" (Weisberg, 2006, p. 739). Weisberg argues that, once it has been determined that a satisfactorily large and varied number of models show the same causal structure and respective robust property, then one can assume that it is likely that the same causal structure is at play when the aforesaid robust property is detected in the real world.

The second question is dealt with in step four of RA, where the question to what extent the robust property is stable under varying background assumptions is addressed.

Weisberg's goal is to show that Orzack and Sober's concern that RA promises a non-empirical kind of confirmation is unfounded. He notes that the third step in his interpretation of the RA-process constitutes the kind of jump from a pure mathematical statement to an empirical one, which Orzack and Sober criticised. However, Weisberg counters that this process "is actually part of a well-accepted theoretical practice that is so common, it is rarely discussed explicitly" (Weisberg, 2006, p. 740). He argues that there is often an

implicit step in the confirmation process before the empirical confirmation, where the scientists ask the question whether, if the causal connection actually holds in the real world, the model would actually be able to represent this process appropriately. Weisberg calls this "low-level confirmation":

> Despite rarely being discussed explicitly, theorists' confidence in their ability to represent phenomena with their models did not come for free. It was minimally established by demonstrating that the relevant mathematics could be deployed to make correct predictions. It may also have been investigated explicitly by mathematicians. These investigations result in what I will call *low-level confirmation*, confirmation of the fact that certain mathematical structures can adequately represent properties of target phenomena. (Weisberg, 2006, p. 740)

Weisberg argues that in his conception of RA low-level confirmation is an element of the third step. It allows us, he argues, to draw a conclusion about the causal relationship between the robust property and the model structure. That is, low-level confirmation is helpful to make the step from a description of a pure mathematical relationship to some empirical assumption.

Herein lies for Weisberg the strength of RA. In those cases where empirical confirmations are difficult due to the complexity of the system, "it identifies hypotheses whose confirmation derives from the low-level confirmation of the mathematical framework in which they are embedded" (Weisberg, 2006, p. 741).

Although Weisberg does not give it a great deal of attention, much of his argument hinges on the condition of a "sufficiently heterogeneous set of models" (2006, p. 739). However, the subsequent discussion (Schupbach, 2018), specifically about the applicability of RA to climate modelling (Lloyd, 2010; Parker, 2011; Winsberg, 2018), centre on the question of what constitutes "sufficiently heterogeneous".

Before I will turn to this debate, I will sketch out what kinds of uncertainties climate scientists face and how climate science deals with this, to see why some philosophers of science have voiced doubt whether RA as outlined above can be applied to climate modelling and if there are alternative routes to establish robust evidence for climate science hypotheses.

3.3.3 Uncertainties in climate science

There are many sources of uncertainties in climate science. This is not to say that there is actually serious doubt in the scientific community that anthropogenic climate change is indeed happening or that its consequences, even with conservative estimates of temperature rise, would not be severe, as every new edition of the IPCC report demonstrates. After all, many relevant processes of the climate and climate change are well understood. Particularly, the link between an increase of carbon dioxide emission and the rise of General Mean Surface Temperature (GMST) has been known about for well over a century (Arrhenius, 1896). Nevertheless, uncertainties play a major role in the public climate-change discourse. So before turning to what makes climate scientists confident in their work despite uncertainties, let us take a look at what causes these uncertainties and how they are communicated.

Broadly speaking sources of uncertainties in climate modelling can be categorised into three different types (Lehner et al., 2020):

1. Model uncertainty
2. Climate variability uncertainty
3. Scenario uncertainty

Model uncertainty is of an epistemic nature (Knutti, 2018, p. 329); that is, not inherent to the system but arises due to our lack of understanding of and lack of means to represent the climate system. These will be explored in more detail below divided into structural and numerical uncertainties, and parameter uncertainty.

On the other hand, uncertainties coming from the internal variability are innate to the climate system. The climate system does not just deviate from the mean state due to external anthropogenic (e.g., greenhouse gases) or natural forcing (e.g., volcanic eruptions) but also internal (e.g., the El Niño-Southern Oscillation) processes. When it comes to assessing climate change with the help of models, internal variability becomes an issue, particular on a shorter time scale, when (disregarding all model uncertainty) it is not clear if a change in the climate is due to external forcing or some random internal variability. These effects can be dealt with by running the model multiple times while

varying the initial conditions[74] and by averaging the relevant climate variables over a longer time period. However, due to the computational costs of making enough model runs to sufficiently explore internal variability it can remain a significant source of uncertainty particularly for shorter timescales.

I will not discuss uncertainties about possible emission scenarios here in any detail. However, they are by no means unimportant. Quite the contrary, as climate models only project how the climate will change under specific emissions scenarios, they are vital to questions of climate-change policy. But these are uncertainties that are directly dependent on human behaviour, not natural laws and lack of knowledge thereof and, therefore, will not be examined here in detail.[75]

Climate models are also not free from the effects of observational uncertainties considering the "symbiotic" nature of the relationship between models and data (Edwards, 2010, 1999). The variety of sources of data uncertainties has been examined in Chapter 3.2.3 and should also be kept in mind in the context of issues concerning model evaluation.

---

74 When it comes to climate projection as compared to predictions, it is commonly said that these are independent of initial conditions as models for projections are not run from observation-based initial conditions but from an assumed preindustrial state. As climate projections are used to explore the impact of external forcing on the climate system in general, the specific initial conditions are less relevant. Internal variability is sometimes nevertheless referred to as *initial-condition uncertainty* in the context of projections – when there is uncertainty regarding the question if the model spread is due to external forcing or 'normal' internal variability because the model performance has not been explored systematically enough in respect to varying initial conditions. However, from a philosophical perspective to what extent it can be said that climate projections are or are not affected by initial conditions uncertainty hinges on the precise definition of the term as Werndl points out. (For internal conditions in the context of predictions and projections, see Werndl, 2019; for the difference of the significance of internal variability to weather forecasting and climate projections, see also Winsberg, 2018).

75 Climate scientists usually consider human behaviour to be external to the climate system. One might speculate whether this has a more fundamental reason that goes beyond pure practical considerations about the models as Parker writes: "Classifying human activities as external to the climate system seems to be a pragmatic choice—it is easier, and a reasonable first approximation, to represent anthropogenic greenhouse gas emissions as exogenous variables in climate models—though it may also reflect a deeper ambivalence about whether humans are part of nature" (Parker, 2018).

3.3.3.1 Numerical approximation and structural uncertainty

No climate model is a perfect and complete copy of the real climate system. There are two reasons for this.[76] On the one hand, climate models are numerical models. At the core of ESMs and similar models are fundamental partial differential equations (see Chapter 2.1). The problem is that these equations cannot be solved analytically. Computer models can only offer numerical approximations, whereby the globe is divided into (digital) grid cells and the equations are solved approximately in discrete time steps. This makes it even theoretically impossible – that is, if one was in a position in which one could integrate every single process of the climate system into the model – to develop a computer model that creates a perfectly digital copy of the climate system.

On the other hand, from a practical point of view, not all parts of the climate system can structurally be represented equally well within a model. When developing a climate model, scientists have to make concessions, in one way or another, regarding which processes are (better) represented and which are not. Idealisations are a necessary part of any climate model. This is not a new development in science. After all, our common definition of model usually implies that it is an idealisation of something (Cartwright, 1983).[77] This is usually not considered a deficit but a clear advantage of models. Whether or not specific idealisations of a model are an asset depends on the particular purpose. For certain (e.g., educational) purposes, the Bohr model of the atom, which envisions the electron to circle the nucleus in a perfect circle, is sufficient. However, when one discusses more complex atom structures than the hydrogen atoms, for instance, to explain the Zeeman effect, one has to make use of other more advanced models. Still, for other purposes atoms need not be considered as more than a point charge, as in the kinetic theory of gases. This does not just hold for the use of models in physics. For example, ecological or economic models are well known for simplifying and reducing very complex systems.

The difference between climate models or many other kinds of computer simulations that deal with added epistemic challenges due to a high complex-

76 Strictly speaking, both aspects, numerical approximation and structural uncertainty, actually do represent two distinct kinds of uncertainties, which, nevertheless, are so interconnected that in practice they are difficult to handle separately (Winsberg, 2018, p. 91).

77 Model *in this context* should be understood in a broad sense (as Giere, 2006 for instance, does), not as a physical object or computer representation of a system but as a specific representation of a more abstract theory.

ity in the target system (see Chapter 2.1) and the examples above is the increasing difficulty to pin down exactly what effects these idealisations have on the model output. What one has to keep in mind is that a 'good' fit with observations cannot necessarily be traced back to a 'truthful' representation of the climate system by the model (see Chapter 3.2.4). It might very well that the good fit is due to compensating effects within the model, following from characteristic features of the model building process such as the fuzzy modularity and the tuning of models. As shown in Chapter 2.1, when assessing the 'quality' of a climate model, scientists face (at least) serious obstacles in gaining "analytical understanding" (Lenhard and Winsberg, 2010, p. 254). This is an issue that also affects the possibility of the application of RA to climate modelling, as will be discussed further below.

All of this makes it harder to pinpoint errors within the model and, in turn, to assess the uncertainties relating to the model structure. One method, as we will see, that climate scientists rely on to explore the structural uncertainty are multi-model ensemble studies (see Chapter 3.3.3.3).

3.3.3.2 Parameter uncertainty

Many processes relevant for the climate system concerning for instance cloud formation, radiation or vegetation growth cannot be resolved directly in the models. Any kind of process taking place on a subgride scale can only be integrated into the model in the form of parametrisations (McFarlane, 2011).

It is often the case that there are different options of how a specific process can be parametrised. Depending on how well the underlying mechanical processes are understood and can be expressed in terms of physical laws, parametrisations can be primarily derived from these and can be akin to a small model within the model with some empirically acquired parameters or based mostly on observationally derived approximations. As with the creation of models themselves, there are usually different options how to parametrise specific processes. In practice, parameter uncertainty is intertwined with structural uncertainty because the choice of a parameter value is very much contingent upon both the resolution of the model and the general model structure. The size of the grid cells will influence optimal parameter values and whether a certain process even necessarily needs to be parametrised. On the other hand, the overall model structure can influence the choice of parameter, for the particular interdependencies within the model might change what the best parameter values is. This is why the process of creating

parametrisation (as well as tuning) is sometimes liked to being an "art form" (Edwards, 1999, p. 445).[78]

Improvements in the resolution can reduce the necessity to parameterise certain processes. However, this usually goes hand in hand with a significant increase in demand of computing power. Many non-negligible climate processes take place on a scale of a few kilometres, meters or way below that, such as those concerning the cloud microphysics. Current state-of-the-art ESMs have a resolution that is still far from that kind of resolution (see Chapter 2.1).

Parametrisations are deeply entrenched in the models. The consequence is that parametrisation schemes are not necessarily replaced or improved within a working model, even when there is a 'better' alternative available. As exchanging parametrisations require careful adjustments and tuning, doing so can be a costly undertaking (Guillemot, 2017).

However, it is not always computing power which limits scientists' abilities to resolve all relevant processes. Certain, specifically small-scale, processes are often not understood well enough to be resolved even if the scale of the models were small enough. As Knutti puts it: "There is simply no fundamental equation to describe how a tree grows" (Knutti, 2018, p. 328).

3.3.3.3 Second-order uncertainty

Why is it so complicated to pinpoint these uncertainties? Could one not just simply compare the model to observations? But as we have seen in Chapter 3.2.3, the available observational datasets themselves are generally extensively processed and come with a variety of uncertainties. Furthermore, one cannot simply interpolate from today's climate to that of the future under anthropogenic forcing. Structural epistemic obstacles such as not fully understood feedback processes make this impossible. Therefore, understanding the models, their strength, their shortcomings and, specifically for the purpose of climate-change projections, how all of this manifests in uncertainty estimates is paramount.

78 A more in-depth discussion of the application of the term *art* to certain methods in climate modelling will following in Chapter 4.2.2.

3.3.3.3.1 Ensemble studies

One way to explore these uncertainties are so-called ensemble studies (see Figure 5). To investigate uncertainties in the model structure, scientists compare different models under a fixed emission scenario with each other in so-called *multi-model ensemble studies* (MIPs). There is variety of model-intercomparison projects for all types of models. The most well-known among these is the Coupled Model Intercomparison Project (CMIP), which has been running since 1995 and is now in its 6th phase. The CMIP assessment of climate model uncertainty is an important contribution to the IPCC assessment reports. The uncertainty of parameter values can be investigated in a similar fashion. In *perturbed physics ensemble studies* (PPE) parameter values are varied in a model within the realm of what scientists consider reasonable to assess how this affects the overall model performance. A prominent example of such a project is done by *climateprediction.net*. In this project scientists have engaged the public so that they offer up free computing power on their private computers to run different versions of the same model (Stainforth et al., 2005). A newer development are *Initial Condition Ensembles* (ICE), which are ensembles based on one model with varying initial states under an otherwise fixed scenario, used to explore simulated internal variability (Chen et al., 2021, p. 222).

Particular MIPs have attracted the attention of philosophers of science, as they at first glance might seem similar to conventional statistical sampling methods but are not. For one, this would usually require sampling independently from the whole space of possible models. But as others (Parker, 2010; Winsberg, 2012) have pointed out, it is hard to imagine how one would even go about doing so. What is more, one would not even want to sample from the whole space of possible models. Introducing models of which we know that they are unrealistic into the uncertainty estimation seems to be contradictory to the purpose (Winsberg, 2012).

In addition climate models are also not truly *independent* of each other in the way it would commonly be required for statistical analysis. As we have already seen in Chapter 3.1.3.2, climate models have shared histories. Climate models are generally not entirely built from scratch but are usually at their core related to other models and contain parts (everything from lines of code to whole parametrisations schemes) also used in other models (Boé, 2018; Knutti et al., 2013).

Figure 5: Illustration of common types of model ensemble, simulating the time evolution of a quantity Q (such as global mean surface temperature)[79]

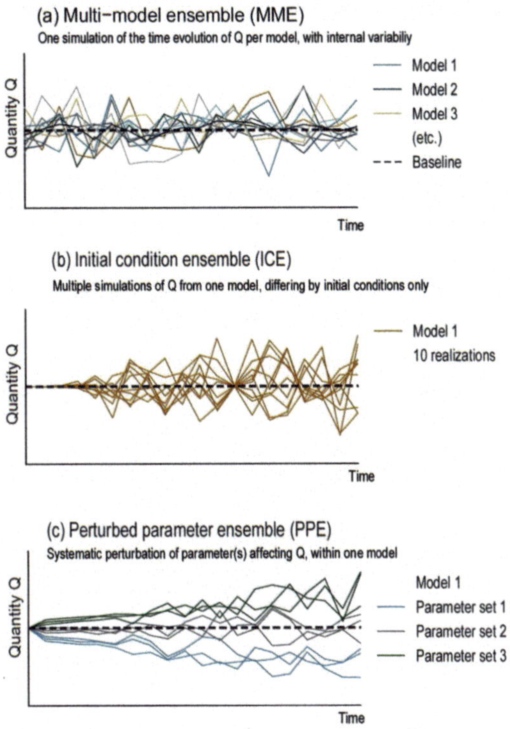

Source: Chen et al., 2021, p. 222, Figure 1.21

79 (a) Multi-model ensemble, where each model has its own realization of the processes affecting Q, and its own internal variability around the baseline value (dashed line). The multi-model mean (black) is commonly taken as the ensemble average. (b) Initial condition ensemble, where several realizations from a single model are compared. These differ only by minute ('micro') perturbations to the initial conditions of the simulation, such that over time, internal variability will progress differently in each ensemble member. (c) Perturbed physics ensemble, which also compares realizations from a single model, but where one or more internal parameters that may affect the simulations of Q are systematically changed to allow for a quantification of the impact of those quantities on the model results. Additionally, each parameter set may be taken as the starting point for an initial condition ensemble. In this figure, each set has three ensemble members.

While MIPs might give the impression that it is an 'objective' approach to quantifying uncertainties, as Winsberg points out (2018, pp. 96–100), it is not an objective method in the sense of complete independence from the expert judgement of the scientists. Therefore, objective here essentially cannot mean anymore than 'not done by hand'.[80]

Traditionally, it is assumed that in an ensemble all models are equal and therefore are weighted the same. But considering that, one the one hand, models are often related to one another and, on the other hand, not all models are equally good at representing all aspects of the climate equally well, the assumption that all models in an ensemble should get the same 'vote' is questionable. Some attempts have indeed been made to weigh models according to performance and independence (e.g., Knutti et al., 2017), but there is no clear consensus among climate scientists on how this ought to be done (Chen et al., 2021, p. 226).

Further, philosophers and sociologists of science have in the past been attesting scientists a "herd mentality" (Winsberg, 2012; see also Sundberg, 2011) when it comes to constructing and evaluating climate models.[81] As the model development is not fully epistemically constricted, it is not uncommon that new models are matched to those that are well established. Social structures within the scientific community work in such a way that modelling groups usually try to avoid being the 'odd one' with a model standing out from the mass. Mikaela Sundberg (2011) argues model agreement creates a kind of "social authority" that scientists follow and adjust new models to in order to be taken seriously.[82]

'Subjective' estimates about the quality of the ensemble study in order to gain a full picture of all uncertainties are a necessary feature of climate-model assessment (Parker, 2014; Winsberg, 2018, pp. 96–102).[83] Nevertheless, ensem-

---

80 Note that understanding objectivity in this way is similar to a definition of objectivity that is actually widespread in climate science, e.g., to describe specific algorithms (see Chapter 2.3 and Chapter 3.4.3).

81 Winsberg actually compares this to Walter A. Shewhart's assessment of historic speed of light measurements, which converged despite being far from the right value (2012, p. 100, see also Shewhart and Deming, 1939).

82 Sundberg argues that this kind of "social authority" does not just affect MIPs in climate science but also intercomparison projects in other fields of science that heavily rely on complex computer simulations such as astrophysics (2011).

83 There is of course also the risk of a kind of herd mentality in more general terms when it comes to expert judgements. The IPCC remarks on this in the *Guidance Note for Lead*

ble studies are considered to be an essential element in exploring structural and parameter uncertainty in climate models and to investigate the effects of climate change. We will come back to this further below.

3.3.3.3.2 The quantification problem

Despite of all these uncertainties, policy makers often expect climate scientists to give clear and precise estimates of how the climate will change under what conditions. This is understandable; after all, in order to tackle the problem of climate change; it is useful to have as clear an understanding as possible of who and what is affected to what extent.

This, however, can potentially put scientists in a difficult situation when they try to stay true to their assessment of uncertainties and still convey helpful and concrete uncertainty estimates. But not providing any uncertainty estimates would not be feasible either as it bears the risk that someone less qualified might feel called to fill the gap.

To tackle this issue the Intergovernmental Panel on Climate Change (IPCC) has created a framework for its authors instructing them on how to communicate uncertainties. The *Guidance Note for Lead Authors of the IPCC Fifth Assessment Report on Consistent Treatment of Uncertainties*, which functioned as a baseline for both AR5 and AR6 (Chen et al., 2021, p. 169), gives the authors two options on how to convey the certainty or uncertainty of their findings:

> The AR5 will rely on two metrics for communicating the degree of certainty in key findings:
> · Confidence in the validity of a finding, based on the type, amount, quality, and consistency of evidence (e.g., mechanistic understanding, theory, data, models, expert judgment) and the degree of agreement. Confidence is expressed qualitatively.
> · Quantified measures of uncertainty in a finding expressed probabilistically (based on statistical analysis of observations or model results, or expert judgment). (Mastrandrea et al., 2010, p. 1)

Authors where the authors are advised to be "aware of a tendency for a group to converge on an expressed view and become overconfident in it" (Mastrandrea et al., 2010, p. 2).

Figure 6: A depiction of evidence and agreement statements and their relationship to confidence. Confidence increases towards the top-right corner as suggested by the increasing strength of shading.

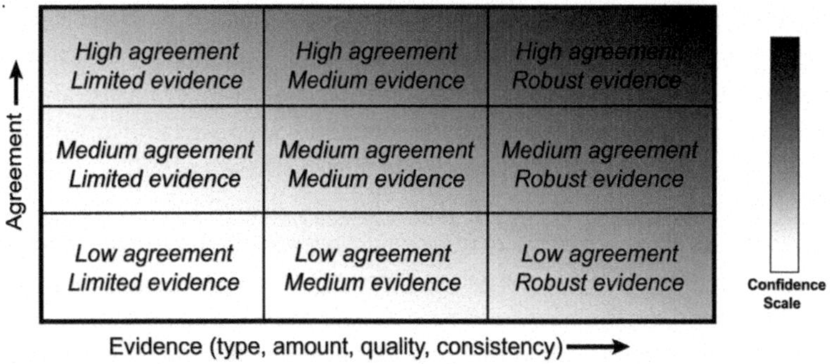

Source: Mastrandrea et al., 2010, p. 3, Figure 1

The IPCC accompanies this with two helpful charts, which further demonstrate how confidence and likelihood should to be understood and communicated (Figure 6 and the likelihood scale already introduced in Chapter 3.1.3.3 Figure 2).

On the one hand, scientists can express confidence in a qualitative way as relating to evidence and agreement. It is stressed in the Guidance Notes that *confidence* should not be interpreted probabilistically. On the other hand, the authors are encouraged when they come to the conclusion that evidence and confidence is sufficient to articulate assumptions about the certainty of specific events or results as *likelihood*. Here, specific terms to convey findings of likelihood (such as *Very Likely*, *Likely*, *Unlikely*) are assigned to margins of outcome probability (90–100 %, 66–100 %, 0–33 %), as shown in Figure 2.

Therefore, the authors of the IPCC assessment report are given a framework that is at the same time calibrated but also somewhat flexible to account for the specificities of a particular hypothesis and the evidence for it. Nevertheless, the *Guidance Note for Lead Authors* also advices the authors of the IPCC report to consider potential pitfalls in how the wording of their findings might be misinterpreted:

> Be aware that the way in which a statement is framed will have an effect on how it is interpreted (e.g., a 10% chance of dying is interpreted more negatively than a 90% chance of surviving). Consider reciprocal statements to

> avoid value-laden interpretations (e.g., report chances both of dying and of surviving). (Mastrandrea et al., 2010)

The IPCC as the UN's organisation that is entrusted with gathering and assessing the current state of climate science is, as Edwards puts it, a unique "hybrid scientific/political organization" (1999, p. 460). Therefore, it has to be particular careful when it comes to communication uncertainties as the reports are not just read by members of the scientific community who are familiar with the conventions but also outsiders who might misread those writings.

However, despite this strict framework, several studies have shown that assessments made according to these guidelines are often misunderstood by laypersons. Particularly wider uncertainty intervals are often misinterpreted to mean that the scientists were less certain than when the interval was smaller (Løhre et al., 2019). How the scientists' assessments are interpreted can also be dependent upon cultural backgrounds (Harris et al., 2013). While AR6 still makes use of the framework given in the Guidance Note originally developed for AR5, the difficulties with communicating uncertainties is also acknowledged there (Chen et al., 2021, p. 171).

After having established, first, the different sources of uncertainty that come along with climate modelling, second, the difficulties in determining how strong these uncertainties are and, third, the language the IPCC applies in communication these uncertainties, the question now remains how the IPCC comes to conclusions, despite all these uncertainties, about the likelihood of (and confidence in) hypotheses about the various effects of increased climate forcing. In order to discuss this question, we will return to RA.

3.3.3.4 Robustness revisited

Returning to RA, the central questions now are: what should be inferred from the fact that projections from different climate models agree? And can we tell if the models agree because of some common essential and true core of the models or because of some specific idealisations of the models? As discussed above, ensembles are not statistical sampling methods. Philosophers and climate scientists generally agree that due to the lack of interdependence of models ensembles cannot be regarded as statistical sampling methods (e.g., Knutti et al., 2017; Parker, 2018; Winsberg, 2018). But is there still some kind of epistemic significance to ensembles of models? The follow-up to this question is: can *more* models increase confidence in their output?

Some philosophers have indeed expressed hope that, in a certain respect, RA might be of help here (Lloyd, 2010, 2009; Winsberg, 2018), while others have been less convinced (Parker, 2011).

In the context of climate modelling RA was first discussed by Elisabeth Lloyd (2009). She argues that Weisberg's (2006) version of model robustness, as represented in Chapter 3.3.2, could also be applied to climate model ensembles. One might, for instance, find in respect to an ensemble of models that

> in all of them [the models] there is a significant role played by greenhouse gases in the late twentieth-century warming of the global climate, and that these are linked to the surface temperature rising in the equations, despite the fact that climate models vary in their assumptions about other aspects of climate. Thus, we would have an analysis isolating greenhouse gases linked to temperature rise (the common structure), and a robust theorem linking greenhouse gases to the robust property, the outcome of rising global mean temperature. (Lloyd, 2009, p. 220)[84]

However, even if all models show a connection between rising temperatures and greenhouse gases, the question still remains how can we be certain that greenhouse gas emission is the relevant factor. To answer this, Lloyd notes, Weisberg makes an "implicit appeal" to a variety of evidence argument:

> he is explicitly appealing to a range of instances of fit of the model over different parameter values, parameter space, or laws. It is against this background of differing model constructions that the core structure occurs and causes the robust property to appear, and it is the degree of this variety of fit for which the model has been verified that determines how confident we should be in the causal connection. (Lloyd, 2010, p. 981)

Lloyd concludes that when a diverse set of models agree so that a robust theorem can be formulated and they also show other instances of fit, we have good reason to be confident in regards to the robust property. Besides a good fit with observations of GMST of the 20th century, different climate models, even

84 The robustness theorem then would be: "Ceteris paribus, if [Greenhouse gases relate in lawlike interaction with the energy budget of the earth] obtains, then [increased global mean temperature] will obtain" (Lloyd, 2010, p. 950).

of a specific type, contain a variety of different background assumptions,[85] parametrisations and parameter values which are, while sometimes contradictory, in themselves empirically supported, Lloyd argues (see also Lloyd 2015). Thus, Lloyd contends that model robustness in this context can have a confirmatory, not just heuristic dimension when not only multiple models with the same causal core converge towards a specific result, but when there are also a variety of diverse but empirically supported modelling assumptions.

Wendy Parker (2011), by contrast, argues that ensembles of climate models on their own do not warrant any conclusions about the truthfulness of or confidence in an ensemble result. Building on Orzack and Sober's argument (1993), Parker reconstructs RA for an ensemble of climate models as follows:

1. It is likely that one of the models in this collection is true.
2. Each of the models in this collection logically entails hypothesis *H*.
It is likely that *H*. (Parker, 2011, p. 583)

Parker notes that such an argument is problematic in the context of scientific models, as idealisations and simplifications are an unavoidable feature. In fact, idealisations are what make models models. Thus, in some way, a model is always false.

However, one can transform the argument in such a way that only the "likely adequacy" (Parker, 2011, p. 584) of a model for a specific purpose is required:

1'. It is likely that at least one simulation in this collection is indicating correctly regarding hypothesis *H*.
2'. Each of the simulations in this collection indicates the truth of *H*.
It is likely that *H*. (Parker, 2011, p. 584)

However, Parker argues that today's climate models neither in respect to the ensemble's performance nor its construction can fulfil the likely-adequacy condition. Concerning the latter, Parker points out that (as already discussed above) ensembles do not sample from the whole space of possible model but are *ensembles of opportunities*. Therefore, it cannot be argued that it is likely that one of the models in an ensemble is indicating correctly regarding *H*

85 Lloyd argues that even models with limited changes in the model structure can be considered sufficiently diverse in the case of climate models, as the non-linearity and feedback process will sufficiently diversify the model output (2015, p. 65)

as the ensemble does not reflect the whole range uncertainty. And as far as model performance is concerned, Parker is also sceptical that a good fit with observations in respect to a particular variable can be rightfully interpreted as an indication that it is likely that one model in the ensemble comes close to predicting the true value of that variable at some point in the future because of the intricate ways of model-data interdependency (see Chapter 3.2).[86]

Eric Winsberg (2018, p. 179) interprets the reason that Lloyd and Parker come to such different conclusions about the applicability of RA to climate science is that they essentially ask two different questions. Lloyd's goal, according to Winsberg, is to explore whether models plus other evidence could support a climate science hypothesis, whereas Parker's approach to RA focuses only on models. Considering these different premises, Winsberg comes to the conclusion, as we will see in the following, that both Parker and Lloyd make valid points concerning RA.

Winsberg begins his analysis of RA (2018, pp. 183–206) by expanding the notion of RA beyond models to a variety of types of evidence coming from a combination of different sources, such as experiments and observations. Further, stemming from the question what is actually meant by "sufficiently diverse" (Weisberg, 2006) for a set of models or/and other evidence to be considered for RA, Winsberg sets out to look for a concept of *RA-diversity* that also "acknowledges that science at best offers grounds for increasing one's degree of belief in a hypothesis" (2018, p. 185). Inspired by Schupach (2018) Winsberg argues that as all climate models share at least some common assumptions (and one would hope so), they cannot be considered to be (fully) independent of one another, one would need a concept of RA-diversity that is not, as is commonly assumed, built on a notion of "probabilistic independence". That is, one would need to find a kind of RA-diversity that also holds for those cases where there is some kind of entanglement either among the pieces of evidence or methodology. Jonah Schupbach (2018) provides such concept of RA-diversity. Studying the application of RA in the context of different scientific practices such

86 Parker further explores whether "increased confidence in H is warranted" or "the security of a claim to have evidence for H is enhanced" (Parker, 2011, p. 581), instead of "likely adequacy", might be more successful approaches. But for similar reasons to those already discussed, Parker argues that these weaker requirements do not suffice to consider RA successful in the case of climate model ensemble studies.

as experiments, observations and models (computer and other kinds), Schupbach asks the question "what is accomplished in successful RAs by introducing diverse means of detection" (2018, pp. 286–287). He answers this question by arguing that these diverse detection methods provide ideally "competing explanations". He proposes a concept of RA-diversity that is characterised by increased confidence in a hypothesis when a new piece of evidence is added. It is defined in the following way:

> Explanatory RA-diversity: Means of detecting *R* are RA diverse with respect to potential explanations (target hypothesis) *H* and its competitors to the extent that their detection (R_1, R_2, ..., R_n) can be put into a sequence for which any member is explanatory discriminating between *H* and some competing explanation(s) not yet ruled out by the prior members of that sequence. (Schupbach, 2018, p. 288)

Based on this, Winsberg calls it the "cumulative epistemic power" (2018, p. 185) of a set of models or other types of evidence when it is rational to assume that every new piece of evidence increases our confidence in a hypothesis.

Appling this concept of RA-diversity to climate modelling, as Winsberg does (2018, pp. 192–193), one might assume a model that 'detects' that equilibrium climate sensitivity (ECS), that is, in short the change of the global mean surface temperature after a doubling of CO_2 in the atmosphere, is above 2 °C. However, the question arises if there are no other explanation for this model result other than that the hypothesis is correct such as some kind of distortion in the model, for example, the particular grid size or the wrong cloud parametrisation. Thus, one has to ask what other detection procedures exist that could rule out these other competing explanations, e.g., by trying out an ensemble of models with different grid sizes or other cloud parametrisations. This process, as Winsberg points out, requires making specific case-dependent decisions about whether or not the competing hypotheses are sufficiently dismissed.

This also means, Winsberg notes, that, whether a set of models is *adequately RA-diverse*, can only be determined with respect to a specific hypothesis and with respect to the question asked and to what extent one can rule out competing explanations. Nor would saying that a set of models and/or other lines of evidence is RA-diverse necessarily be a statement on whether or not to ac-

cept a hypothesis. It is only a sign for cumulative epistemic power, concludes Winsberg, that is, being on the right track.[87]

Though RA defined in this way is applicable to all sorts of lines of evidence, it may also be applied to just models. Whether or not an ensemble of models is considered RA-diverse is only a question of the particular hypothesis, argues Winsberg: "You can make an ensemble of opportunity RA diverse without altering the ensemble by altering the hypothesis" (Winsberg, 2018, p. 202).

Further, establishing explanatory RA-diversity will frequently require an understanding of the underlying structure of the model and the climate system in general with respect to the relevant processes, as we will see in the following.[88]

As an example of how this definition of RA-reasoning can be applied to climate-change hypotheses, Winsberg examines in a case study the problems of establishing a value for ECS (2018, pp. 194–206). He particularly focuses on an estimate from CMPI5 of ECS being between 2.1 °C and 4.7 °C, as reported in AR5 (Flato et al., 2013, p. 818).[89] The difficulty in estimating ECS can be attributed to the fact that it is not just the outcome of increased CO_2 in the atmosphere but also several feedback processes, with cloud feedbacks considered to be the biggest issue.

Feedbacks are also at the root of why Winsberg is sceptical that ensembles of opportunity (on their own) can provide robust evidence, that is, it cannot be demonstrated that the ensemble is sufficiently RA-diverse when it comes an estimate of ECS in the sense of the hypothesis above. Let's see why. First of all,

87 It is important to stress that *RA-diverse* is different from *sufficiently diverse* insofar as RA-diverse does only imply that a model set "gets better as it gets larger" (Winsberg, 2018, p. 186). Whether or not to accept a specific hypothesis is, then, context-dependent and among other things a question of inductive risks (see Chapter 3.1).

88 Winsberg links this to Knutti's definition of process understanding (Knutti, 2018, see also Chapter 3.2.4). In fact, Winsberg claims "gaining process understanding is not necessarily a separate kind of epistemic activity from RA, and the two are complementary, rather than competing, accounts of how we gain confidence in model results" (Winsberg, 2018, p. 202).

89 While the latest IPCC report (AR6) does no longer directly consider models for assessing ECS, it is nevertheless helpful for the questions asked here, that is, how RA can be applied to models as well as other sources of evidence, to take a look at Winsberg's appraisal of the reasoning process behind the AR5 assessment of ECS, even if it is somewhat out-dated.

taking into account all we know about the challenges of creating ESM, all models agreeing on a particular estimated range of ECS could still be an "artefact of the systematic failure of all the models to accurately capture all of the feedbacks – with cloud feedbacks being an especially likely candidate" (Winsberg, 2018, p. 196). Further, even if the models fit well with the available observational data of the relevant feedback process, this does not necessarily mean – as we have seen in Chapter 3.2.4 – that the models would also necessarily adequately capture the effect of feedbacks for a possible future climate that differs significantly from what we currently have observations for. It might as well be that the models only fit so well with the available data because of some compensating errors in the models that will cease to compensate in the same way under future climate change conditions. Thus, considering the significance of cloud feedbacks to assessing ECS, Winsberg concludes that for the models of an ensemble to be RA-diverse regarding a hypothesis about ECS, they would have to not just accurately model the cloud feedbacks with respect to the observable past; one would also have to eliminate the possibility of the model fitting well with the data due to error compensation. Only if this were accomplished, argues Winsberg, then "there is a high probability that we are correctly detecting a hypothesis about cloud feedback in the future climate" (2018, p. 197).

In this context Winsberg also points out a more recent trend in climate-change assessment to tackle ECS and similar problems: trying to find so called *emergent constraints* (Chen et al., 2021, p. 225; Winsberg, 2018, pp. 197–201). The goal of this method is to reduce uncertainties in climate-change projections by establishing a relationship between a future climate-change response and present-day observations. Let us assume we are interested in a variable *b* (also called the predictand), e.g., the intensity of a specific feedback process that models in the ensemble projection do not agree on. However, we suspect that there is a correlating relationship between variable *b* and another variable *a* (also called predictor) of a process that is taking place on timescale for which there are good observations. Such a relationship might, for instance, be the snow-albedo feedback (that is, the increase in absorption of solar radiation due to the increase of ice melting because the surface is warming) which is taking place both in a seasonal cycle and under (longer term) forcing conditions (Hall and Qu, 2006). This relationship is then established by running the ensemble once over a short time period to determine the model spread of variable *a* and once under a long-term forcing scenario to determine the model spread of variable *b*. The model output for both variables is plotted, with variable *a* on the x-axis

and variable *b* on the y-axis, and every point on the graph representing one model (see Figure 7).[90] An *emergent constraint* is considered to be "trustworthy" (Caldwell et al., 2018) or "confirmed" (Hall et al., 2019) when it can be further argued that the mechanical process behind the correlation of *a* and *b* is well understood and the emergent constraint is tested 'out of sample' with an ensemble that was not used in finding the emergent constraint, so as to rule out that the emergent relationship is not just a coincidence, as a result of compensating errors and lack of diversity of models in the ensemble (Hall et al., 2019).[91]

Winsberg argues that the reasoning process behind emergent constraints can be considered "one of the best RA reasoning [...] in climate science" (2018, p. 197) to the extent that it is a way to systematically rule out alternative explanations for a hypothesis by establishing and testing the stability of the particular underlying process (2018, p. 201).

While this has been done successfully, e.g., in the case of the snow-albedo feedback (Hall and Qu, 2006), this kind of reasoning is much more difficult to establish when it comes to ECS, which is affected by many different kinds of feedback processes. Thus, returning back to the question if an ensemble of models can be considered explanatory RA-diverse concerning a hypothesis about ECS, Winsberg argues that this requires that it is first demonstrated sufficiently that the ensemble is exploratory RA-diverse in the sense of an emergent-constraint reasoning process in respect to every single feedback effect that gives rise to ECS.

That is why Winsberg is sceptical that a robust reasoning concerning a hypothesis about ECS could be arrived at based on an ensemble of models alone (2018, p. 199). However, Winsberg's definition of explanatory RA is not restricted to models. Thus, if one takes into account that scientists also have access to other detection methods in the form of instrumental records and proxy paleoclimate data as independent lines of evidence then one can see how the authors of AR5 nevertheless come to the following conclusion:

90 This, of course, requires that the range of observations is within the interval of the model spread for variable *a*, so that variable *b* can actually be constrained.

91 If all models in the ensemble are related to each other too much, then the suspicion that the emergent relationship might be an artefact of model error is more pressing than when the models are more independent (Brient, 2020).

Figure 7: The principle of emergent constraints.[92]

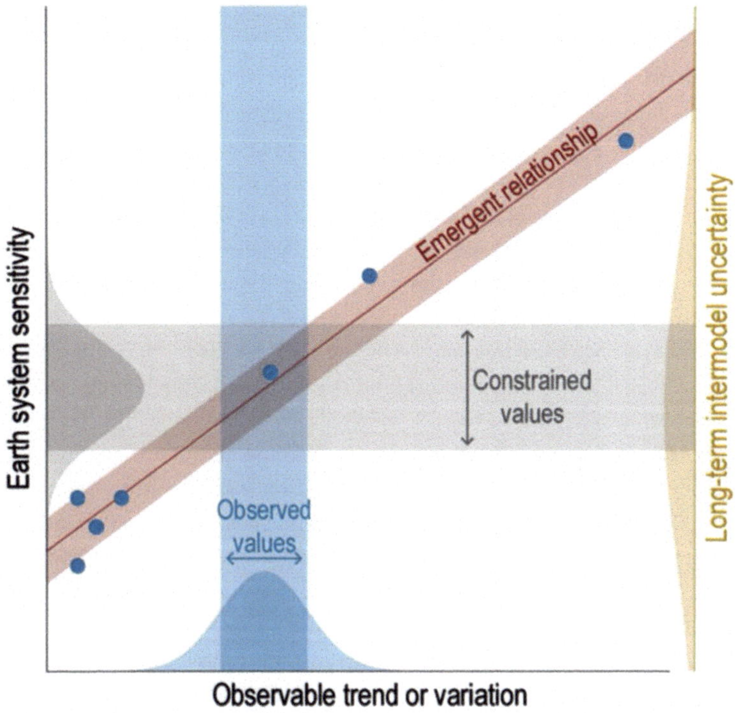

Source: Chen et al., 2021, p. 225, Figure 1.23

Based on the combined evidence from observed climate change including the observed 20th century warming, climate models, feed-back analysis and

92 An ensemble of models (blue dots) defines a relationship between an observable mean, trend or variation in the climate (x-axis) and an uncertain projection, climate sensitivity or feedback (y-axis). An observation of the x-axis variable can then be combined with the model-derived relationship to provide a tighter estimate of the climate projection, sensitivity or feedback on the y-axis. Figure adapted from Eyring et al. (2019).

> paleoclimate, ECS is *likely* in the range 1.5°C to 4.5°C with *high confidence*. The combined evidence increases the confidence in this final assessment compared to that based on the observed warming and paleoclimate only. ECS is positive, *extremely unlikely* less than 1°C (*high confidence*), and *very unlikely* greater than 6°C (*medium confidence*). (Collins et al., 2013, p. 1111)

This statement is based on a set of detection methods that is RA diverse, Winsberg argues, for the following reason (2018, pp. 202–206): models, instrumental records and paeleoclimate data applied as detection methods for ECS all come along with specific uncertainties. However, if not all lines of evidence are susceptible to the same uncertainties (or at least not to the same degree) and one treats "each of these sources of uncertainty [...] as alternative possible explanations of various hypotheses detections" (Winsberg, 2018, p. 205), then it is possible to see how scientists can rule out different alternative explanations one after another and provide robust evidence for a likely range of ECS as well as the possible upper and lower limits, Winsberg concludes.

For instance, paleoclimate data is not much prone to errors coming from internal variations but most probably will suffer from higher measurement uncertainty and might rest on a different base state as many millions of years ago the climate might have been quite different to our current. The latter does not affect instrumental records and when the hypothesis is broad enough (as, e.g., that ECS is between 1.5 °C and 4.5 °C, see above) measurement uncertainties are also less of a concern here. Both types of detection methods are also less susceptible to model errors, as simulations of the type of ESM are, and so on.[93] This kind of reasoning is at the heart of the statement about ECS in AR5, Winsberg argues.[94]

---

93 Winsberg also notes that RA reasoning is also helpful to understand why it is so difficult to constrict ECS at the upper end. Considering the question if there are alternative explanations for why the current detection methods do not project a higher value, there could, for instance, be a yet unknown feedback process, which instrumental records do not yet detect. Winsberg concludes, that we "would probably only expect to see it in the millions-of-year-scale paleodata – but those data sets have enough uncertainty that they are poor at eliminating such a hypothesis" (2018, pp. 206), which makes constricting ECS at the upper end so difficult.

94 For a more in-depth analysis of the argument of the applicability of the idea of RA to the reasoning about the value of ECS, see Winsberg (2018, pp. 203–206). He follows the scientific argument made in Knutti and Hegerl (2008).

In this context it is worthwhile to mention that compared to the assessment of AR5 the authors of AR6 actually do consider *emergent constraints*, alongside *instrumental records*, *paleoclimates* and *understanding of the climate processes* as lines of evidence for the following assessment of ECS:

> Based on multiple lines of evidence, the very likely range of equilibrium climate sensitivity is between 2°C (high confidence) and 5°C (medium confidence). The AR6 assessed best estimate is 3°C with a likely range of 2.5°C to 4°C (high confidence), compared to 1.5°C to 4.5°C in AR5, which did not provide a best estimate. (IPCC, 2021b)

With respect to the discussion above a few things are noteworthy here. First of all, while emergent constraints are now considered a line of evidence, they are not the only line of evidence.[95] The authors particularly point out that emergent constraints in AR6 are not combined "to provide very strong evidence on ECS" because there are still cross-dependencies between different emergent constraints and it is still too new a technique to rule out that there may not be unaccounted systematic biases (Forster et al., 2021, p. 1005).

Secondly, models are no longer a direct line of evidence and find their way into the assessment only indirectly (for instance, through emergent constraints and because they inform process understanding of feedbacks). The authors cite issues pertaining to the specifics of climate-model construction discussed in this chapter, such as lack of model independency, analytical intractability and the difficulties of evaluating and weighing models adequately as reasons for this (Forster et al., 2021, pp. 1007–1009).

Last but not least, the authors argue that one reason for the improvements in the assessment of ECS from AR5 to AR6 was the application of a new structured (Bayesian) approach of taking different lines of evidence[96] into account

95 In respect to Winsberg's argument it is interesting to note that authors of AR6 differentiate between two kinds of emergent constraints on ECS: "(i) those that are based on global or near-global indices, such as global surface temperature and the TOA energy budget; and (ii) those that are more focussed on physical processes, such as the fidelity of phenomena related to low-level cloud feedbacks or present-day climate biases" (Forster et al., 2021, pp. 1004–1005). Only the first kind of emergent constraints are taken into account in the assessment of ECS with the authors citing concerns about possible biases.

96 Sherwood at al. (2020) base their assessment only on three lines of evidence: instrumental record, paleoclimate data and process understanding. Emergent constraints are not considered as a distinct line of evidence.

as done by Sherwood et al. (2020). Sherwood et al. note that the broad agreement of lines of evidence works as "mutual reinforcement" (2020, pp. 73–74; see also Forster et al., 2021, p. 993). They apply what Stevens et al. (2016) have called a storyline-approach to combine different lines of evidence by laying "out all the circumstances that would have to hold for the climate sensitivity to be very low or high given all the evidence" (Sherwood et al., 2020, p. 2). This shows some similarities to Winsberg's concept of RA in the sense that it is not just an appraisal of models but different types of evidence and more importantly, that it is a systematic approach to rule out alternatives (though in this case alternative storylines, not alternative explanations for a hypothesis).[97]

3.3.4 Conclusion

There are, as we have seen in this chapter, considerable obstacles to be navigated in order for climate scientists to make any kind of assumption about the future of the climate. However, while the sources of uncertainty are manifold and often not easy to assess or minimise, this does not mean that the uncertainties are so overwhelming that no conclusion about the anthropogenic climate change can be made. Quite to the contrary, many of the essential variables and processes that determine how the climate changes are well understood. Take, for example, the case of ECS discussed in this chapter. While the estimate of the range has been refined since the assessment in the famous so called *Charney Report* in 1979, the estimate that it lies between 1.5 °C and 4 °C has been consistently confirmed (Charney et al., 1979; Forster et al., 2021, pp. 1006–1007).

In public debates about climate change it has often been argued that the models are not good enough, that the models disagree with the data or that there is still to much uncertainty. But a central insight from Winsberg's analysis of the applicability of RA in climate science that I think is important to highlight here is that the question of whether or not an ensemble of models is robust is the wrong question to ask. Instead the question ought to be whether or not a specific hypothesis can be supported or refuted with the help of very hypothesis-specific detection methods. More broadly speaking, the question

97 For another similarity note that concerning the unavoidability of expert judgement in the whole process Sherwood et al. also point out "solid *qualitative* understanding of how the evidence stacks up is at least as important as any probabilities we assign" (2020, p. 73).

whether or not there are good reasons for accepting or rejecting a specific hypothesis about climate change is usually not a question of model or data or even models versus data but a elaborated reasoning process based on a combination of different lines of evidence.

The argument that has been made here is that, despite the complexity of the climate system and all epistemic challenges, many important questions concerning the future of the climate can be dealt with with careful reasoning. But what climate science (and science in general, for that matter) cannot provide are fully uncertainty-free answers to every question in an irrefutable way and without any doubt. This sometimes clash with public expectation. However, when scientists are faced with demands to only make yes-or-no-statements giving in to these demands does not just misrepresent the scale of scientific knowledge accumulated but also bears the risk of delaying taking action to mitigate climate change, as Isaac points out:

> Typically, policy-relevant issues are publicly discussed in binary terms [...] yet the relevant science is more appropriately framed in terms of degrees of certainty or evidential support [...] A public rhetoric of bivalence obscures the nature of the scientific contribution to our knowledge of the world and undermines its effective use in policy choice. (Isaac, 2014, p. 43)

In public discourse the demand for 'better' science is often framed as a necessity in order to assess whether or not costly mitigation measures actually have to be taken. In practice, though, this has often been misused as a stalling tactic by particular interest groups to advocate against policies which would be to their disadvantage (Howe, 2014; Oreskes and Conway, 2010). Heymann et al. (2017b) interpret this attitude towards uncertainties in science as a consequence of a view of science which emphasises prediction making as a major trait of science:

> As a consequence of their political, cultural, and economic status and value, tremendous resources flow towards the establishment and operation of cultures of prediction. These investments do not always serve the support or justification of decision making and politics, but can also serve to delay or replace decision making and politics – particularly in the case of contested issues with strong inherent political risks. A commonplace argument for the replacement of effective politics is the call for further research, for example, due to perceived or alleged uncertainty and the lack of sufficient knowledge apparently required to make strident decisions. (Heymann et al., 2017b, p. 8)

Thus, the 'sound science'-argument can quickly turn into a 'throwing out the baby with the bathwater' type of reasoning as it neglects the many questions which can be answered with a broad consensus among climate scientists: chiefly amongst them that anthropogenic climate change is happening. But the knowledge that scientists have accumulated about climate change goes way beyond that. Nevertheless, estimating uncertainties is an intricate process where many assumptions have to be made and lines of evidence have to be weighed and examined requiring carful reasoning.[98]

3.4 Looking back and a tentative look forward

This chapter has shown why three specific widespread ideals about how science does and should operate are inadequate to describe actual scientific practice, particularly when there are, like in climate science, additional epistemic challenges due to the high complexity of the target system. All three ideals have a history of being upheld as signs for good, reliable, adequate science. Science being, firstly, a value-free enterprise that, secondly, works with theories which are easily and unambiguously assessable with the help of experiments or observations and, thirdly, that provides clear binary predictions about the future is commonly considered to be hallmarks of good science. In the past, following these ideals has often been considered to be what makes science special and distinguishes science from other human endeavours.

In the context of science dealing with highly complex systems it becomes apparent what has always been the case: science cannot, has never and, most importantly, does not have to live up to these ideals. But what does the inadequacy of these ideals mean for our ability, specifically from the perspective of an outsider to the scientific community, to draw conclusions about the quality or credibility of the methods and hypotheses of, for example, climate science? Is there another way to ground our trust in science? Fortunately, I think there is. It requires shifting the focus from the characteristics of either the individual

---

98 The IPCC here functions specifically as an organisation that screens and assesses all existing research. It also provides policymakers and the general public with summaries and estimates of uncertainties. It should be pointed out that the IPCC is an unusual institution that has only few equivalents in other sciences. The existence of the IPCC is not just a sign of the relevance of climate research for society but also of the complexity of the system as well as the overall issue.

scientist or the research on to the social structures of science. This reasoning and what follows from this will be further explored in the next chapter.

Here, however, I want to first take a quick look back. In Chapter 2, I have introduced three different 'recurring themes' as preliminary remarks: the *epistemic challenges of highly complex systems*, the *DJ distinction* and *scientific objectivity*. I stated that theses concepts would reappear in the in our examination of the three ideals. Now I will revisit these concepts and see how they pan out, specifically in the context of climate science. We will see that these themes were not just contributing factors to the development and implementation of the ideals and that they are a turn of conflict for these ideals with actual science practice but also that the issues with these concepts will also give a first indication of a way out of this dilemma.

3.4.1 Complexity and understanding

Chapter 2.1 gave an introduction to the complexity of the climate system and how scientists navigate this complexity with help of computer simulations. One concern which was raised there is that modelers will inevitably have to deal with some degree of "epistemic opacity" (Baumberger et al., 2017; Humphrey, 2004) or "analytical intractability" (Lenhard and Winsberg, 2010) due to the complexity of the system and models. One question that follows from this is what this means for our ability to achieve *understanding*. While it has traditionally been a primary aim and motivation for scientists to understand 'why the things are the way they are' or 'how things work', Johannes Lenhard (2020) argues that computer simulations have shifted our perception of what science can and cannot accomplish in terms of acquiring *understanding*. He comes to the conclusion that, while complex climate models, like ESMs, are employed by climate scientists to explore and gain knowledge about the global climate system, they also contribute to a reduction in access to understanding due to their high complexity. On the one hand, many essential aspects of the model performance derive from the physical principles and fundamental equations that the model is based on. On the other hand, the models consist, of course, of much more than just these basic equations. ESMs display an intricate relationship between fundamental equations, semi-empirical parametrisations and tuning that is further intertwined through the iterative process of model construction and improvement (see Chapter 3.2.3.3). Because of this Lenhard argues that the conventional strategies for acquiring understanding in science by dealing with the complexity through "stripping off aspects until only the

essence remains" (Lenhard, 2020, p. 2) cannot be applied. This is what Lenhard refers to as the "dilemma of growth" (2020, p. 2); While increasing complexity is a necessary feature of the models to live up to the high complexity of the climate system, the high complexity of the models limits the extent to which it is possible to reach the aim of "getting to the essence of the mechanism" (Manabe, 2006; as quoted in Lenhard, 2020, p. 2).[99]

Nevertheless, Lenhard observes that climate scientists often develop a "feeling" for the model and its behaviour. Scientists regularly rely on the experience they have with the model and derive some informal knowledge of how certain adjustments will most likely affect the performance of the model in question even though the exact inner-model processes that generate this model behaviour are not fully transparent. Inspired by Max Weber's concept of *verstehendes Erklären* (understanding-explanation (1913))[100], Lenhard argues that this is a way to circumvent what he calls elsewhere the "complexity barrier" (2019) preventing scientists from going the more established route to understanding:

---

99 Lenhard notes that one possible strategy to deal with the "dilemma of growth" might turn out to be resorting to a 'hierarchy of models': "The hope is that small and well-understood models can be knitted together in larger hierarchies so that understanding extends to the whole. However, the prospects of this approach are not yet clear, in part because modularity tends to erode in larger simulation models, leading to a problem of 'holism'" (Lenhard, 2020, p. 2). In respect to the problem of finding a 'hierarchy of models' see also Held (2005).

100 In this, as Lenhard argues, two different approaches to conceptualising *understanding* in philosophy of science are combined. Both centre on the terms *explanation* and *understanding* but define the relationship differently. On the one hand, there is the notion that understanding can be gained once an *explanation* in form of a (logical) *derivation* from basic principles is reached. On the other hand, there is the view – coming from a hermeneutical perspective – that sees explanation and understanding as belonging to two different research fields. Explanation is what can be achieved in science. Understanding, though, belongs to the humanities, where understanding is reached when something, or rather someone, behaves in a way that *matches* what one has anticipated.
Lenhard argues that both *derivation* and *match* can be found in Weber's conception of *verstehendes Erklären* and in the kind of understanding that is reached in climate modelling. The equivalent to *matching* in climate modelling is the "feeling" that Lenhard remarks climate scientists develop for their models, but to some degree, there is also an element of *derivation* because at its core the models are, despite semi-empirical parametrisations and tuning, still built on some fundamental equations.

> Such acquaintance with model behavior can be a work-around for building an adequate "inner" representation [...] when simplification/idealization strategies are *not* available—as in the case of ESM. However, this work-around does not lead to understanding in the traditional sense. There are no simple models involved that would enable understanding by "capturing the essence of a phenomenon" (Held, 2005, 1609). Nevertheless, simulation provides understanding—if only in the weaker, pragmatic sense of getting acquainted with model behavior. (Lenhard, 2020, p. 3)

Lenhard contrasts this with a more conventional conception of *explaining* and *understanding* that is based fully on "analytic derivation from first principles" (Lenhard, 2020, p. 3), which in the context of climate modelling (at least at this point)[101] cannot be accomplished. However, Lenhard argues, what can be achieved is a pragmatic conception of understanding that might function as a substitute so long as no other, more satisfying, conceptualisation of understanding can realised (see also van Fraassen, 1980).

3.4.2 Discovery and justification

Another concept introduced in Chapter 2.2 was the distinction between *context of discovery* and *context of justification*. The DJ-distinction was of relevance both for the emergence of the value-free ideal and the theory-centred view of philosophy of science.

Although the aftermath of the brief popularity of the distinction at the middle of the last century can still be felt (see Chapter 2.2), it is a conception that does not seem to have gained much traction in current philosophy of science. For example, in the edited collection *Revisiting Discovery and Justification* edited by Jutta Schickore and Friedrich Steinle (2006b), none of the authors argues to uphold the distinction as a dichotomy in any strong interpretation of the concept. On the contrary, many claim that a context distinction going beyond a weak form in the sense of Hoyningen-Huene's differentiation between a normative and descriptive view of science cannot be maintained (see Chapter 2.2). Specifically, the claim that justification is "'the other' or 'the opposite' of theory construction, experimentation or indeed discovery" is rejected by theses authors (Schickore and Steinle, 2006a, p. xiii). On the contrary, Schickore and

101 It should be pointed out here that Lenhard notes that his argument is not that this concept of pragmatic understanding "should or in fact must become the goal" in climate science (2020, p. 3).

Steinle note in the introduction of the aforementioned book that "[d]iscovery, in any meaningful understanding of the concept, is a prolonged activity that involves both the generation and fixation of knowledge claims" (2006a, p. xiii). In this sense, e.g., Steinle (2006) and Arabatzis (2006) argue that discovery and justification in scientific practice go hand in hand. Steinle argues that this is, especially, the case in the event of *exploratory experiments*:

> Exploratory experimentation is concerned with developing regularities and appropriate concepts. If it is successful, this success consists in formulating ever more general laws. One may well ask whether such laws have then been discovered or justified: after all, in common language we often speak of laws as having been "discovered" by Galileo, Boyle, Hooke, or Mariotte, for example. As soon as we try to clarify our concepts, however, such talk immediately becomes inappropriate: at the moment when laws are formulated in the research process, they are discovered and justified at the same time. Even if a researcher had initially just a speculation of a possible empirical law, she would conceive this law as being "discovered" only in the moment when it was fully supported, i.e., justified. (Steinle, 2006, p. 187)

The notion that we usually only talk of something (such as a theory, phenomena, object) as being 'discovered', when the belief in it is also considered to be justified by the particular scientific community, is also stated by Arabatzis:

> A mere hypothesis to the effect that a new entity exists would not qualify as a discovery of that entity. The justification of that hypothesis would be a constitutive characteristic of that discovery. The context of discovery is "laden" with the context of justification because "discovery" is a term which refers to an epistemic achievement: if one succeeds in discovering something then, no doubt, this something exists. (Arabatzis, 2006, p. 217)

For instance, as Arabatzis (2006, p. 217) points out, these days one would hardly claim that phlogiston was discovered in the 18th century even though at the time many scholars regarded it to be a substantial discovery in chemistry.

Arabatzis argues that there is a distinction to be made between *discovery* and *generation* or *construction*. The difference, he argues, between these terms is their relationship to *truth*. While generation and construction might lead to truth, discovery already "implies truth" (2006, p. 218), which, Arabatzis concludes, makes discovery an "extended process, which involves both generation and justification" (Arabatzis, 2006, p. 226).

However, this also leads many authors of the above-mentioned volume to the conclusion that the DJ distinction can still be upheld in the sense that there is a difference between "original historical model of hypothesis generation and the 'final' form of justification" (Arabatzis, 2006, p. 218). Steinle, specifically, notes that though in practice scientists are very well "aware of the historical nature of their enterprise" (2006, p. 189), retroactively, scientists generally remove, as much as possible, any kind of reference to the process once it comes to communicating research achievements to others scientists and the wider public, such as in journal articles or textbooks – a process also noted by Reichenbach himself (1938, p. 6). This "process of decontextualizing" is also an attempt to portray the obtained knowledge as secure as possible, "i.e., stripped from the specific time, place, and process by which it has be generated" (Steinle, 2006, p. 189).[102] In this sense, Steinle argues, is the DJ distinction part of science itself.[103]

When it comes to climate science which substantially relies on highly complex computer simulations, the notion that a theory or hypothesis can be validated (at some point) independently from understanding the circumstances under which it was generated is not so easy to sustain. As shown at several stages throughout this chapter, the behaviour of a model is often intricately dependent upon its history. Understanding the history of the models is, therefore, constitutive to knowing to what extent a 'good' model output is actually rooted in a good representation of the relevant process in the model or just an artefact of some unknown interference between different elements of the model or some effect from tuning. However, this is easier said than done. As

102 Steinle (2006, pp. 190–193), however, also remarks that the wider conceptual framework of a theory cannot be seen as independent of its historical origin. While these overall concepts (an example from history of science is the concept of absolute space in physics) are unconsciously followed in daily scientific practice, they are also broadly contingent upon the (social) context in which they were established. Nevertheless, depending on the point of view, Steinle argues, a theory can be considered to be justified separately from the context of its generation (that is, when the contextual framework is not taken into account).

103 Steinle sees a form of Hoyningen-Huene's lean version in this insofar as scientists are posing two different questions: "how did a certain insight (a theory, law, fact, ...) come about? And why should we believe it, what are the reasons for support?" (Steinle, 2006, p. 188) once they turn from the research process to communicating and teaching their findings.

we have seen in this chapter, global climate models have some degree of opacity to them, which entails that conventional methods of justification such as a 'simple' fit to observations or more complex ones such as V&V only have limited applicability.

On the other hand, Winsberg (2018, p. 160, 2003) argues, with reference to Hacking (1983), that the techniques applied in complex computer simulations,[104] as used in climate science, have "a life of their own"; meaning that they are justified not just by theory alone but also because they are understood to be well-established procedures:

> Whenever these techniques and assumptions are employed successfully, that is, whenever they produce results that fit well into the web of our previously accepted data, our observations, the results of our paper and pencil analyses, and our physical intuitions, whenever they make successful predictions or produce engineering accomplishments, their credibility as reliable techniques or reasonable assumptions grows.
>
> That is, the next time simulationists build a model, the credibility of that model comes not only from the credentials supplied to it by the governing theory, but also from the antecedently established credentials of the model building techniques developed over an extended tradition of employment. (Winsberg, 2003, p. 122)

Thus, in Hacking's (1983) words, these techniques are "self-vindicating" or, as Winsberg puts it, "they carry their own credentials" (2003, p. 121). That is, similar to Hacking's claim about experiments, the credibility of these techniques lies, according to Winsberg – at least to some parts – in their historical successful application.

In practice, a certain knowledge about the history of the model can also be beneficial insofar as it can reduce the opacity computer simulations of this

104 Winsberg specifies that by "techniques" he is "referring to the whole host of activities, practices, and assumptions that go into carrying out a simulation. This includes assumptions about what parameters to include or neglect, rules of thumb about how to overcome computational difficulties—what model assumptions to use, what differencing scheme to employ, what symmetries to exploit—graphical techniques for visualizing data, and techniques for comparing and calibrating simulation results to known experimental and observational data" (Winsberg, 2003, pp. 121–122).

type bring along.[105] If within a community of modellers the knowledge about some past modelling decisions gets lost over time, it can lead to problems in the future when some new adjustment results in a seemingly unexplainable change in the model's behaviour. There might be a situation, where a haphazardly constructed model implementation due to limited computing power is introduced to the model at a certain moment in time but forgotten after some generation of scientists as it has not interfered negatively with any other model adjustments in the meantime. Lack of knowledge of the history of the model can become a significant obstacle for scientists when a new improvement to a model component all of a sudden does interfere with the previous imperfect and forgotten adjustment and leads to an unexpected bad model performance (Lenhard, 2018, pp. 839–840 describes an example of such a case). If one sees models as instruments applied to gain knowledge about the climate system, as many climate scientists do (Chen et al., 2021, p. 215), then knowledge about the history of the model can strengthening the confidence in the instrument.

All of this makes it questionable if the notion that the procedure of justification can be fully uncoupled from the historical circumstance is still viable. It also shows how the expertise and experience that scientists have acquired throughout working with these models and the techniques employed to develop them are key to assessing the models and resulting hypotheses.

3.4.3 Scientific objectivity

The third concept that was pointed out in Chapter 2.3 was *scientific objectivity*. The ideal of value-free science is often closely connected with the concept of 'objective' science. The problems of this interpretation of *objectivity* were examined in Chapter 3.1. It was shown that a definition of *scientific objectivity* as complete value-freeness of science cannot be maintained in actual scientific practice. An alternative definition of scientific objectivity that is derived from a plurality of perspectives has been shown to have more chances of being successful (Leuschner, 2012a; Longino, 1990). Another related application of the terms objective and subjective we have seen in this chapter refers to the fact that no

105 That disclosing the specific circumstances and methods of model development can improve model evaluations has also been raised by scientists themselves. For instance, Mauritsen et al. (2012) advocate for making the tuning process more explicit in inner-scientific discussions and communications.

perfect model exists and some subjective decisions have to be made concerning which and how specific processes are included (Tebaldi and Knutti, 2007). A further use of the term objective we came across in this chapter concerns a simplified description of the relationship between observations and theories, where observations supposedly provide 'objective', irrefutable evidence that a theory (or model) is right or wrong.

Here however, I would like to take another look at a somewhat more specific conception of *scientific objectivity*, which is more common in climate science. Climate scientists particularly evoke the term *objectivity* when describing mathematical or automatic procedures. *Subjective* approaches are marked by being 'done by hand' and relying on expert judgement. This distinction of objectivity and subjectivity for instance is often used in descriptions of the tuning process (see also Chapter 4.2.2). By objective method "one means that a well-founded mathematical or statistical framework is used to perform the model tuning, for instance, by defining and minimizing a cost function or by introducing a Bayesian formulation of the calibration problem" (Hourdin et al., 2017, p. 594). Contrary to this, the more common approach to tuning is described by climate scientists as "subjective" and is more directly guided by the expert judgement of the scientists. However, as scientists note, application of the afore mentioned objective procedure still has subjective components to the extent that "[a]ny such objective tuning algorithm requires a subjective choice of a cost function and this involves weighting trade-offs against one another" (Mauritsen et al., 2012, p. 16).

Although the scientists note that subjective judgements (at this moment) are unavoidable, it is also discernible that *objectivity* for the scientist has a positive connotation and is to be preferred to subjective methods. The hope put into objective procedures is to find a way to exclude personal influence on research. Subjective methods imply that the scientist has to make some kind of judgement. To make these judgements scientists have to rely on their expertise and experience. But the worry is that this inevitable reliance on this expertise-led decision-making will, considering the general complexity of climate modelling further "an unfortunate reduction in transparency" (Schmidt et al., 2017, p. 3208). Appealing to objective methods is linked to the wish to bring transparency, traceability and reproducibility into these processes.

3.4.4 Conclusion: what now?

What these three concepts have in common that they allude to the relevance of another element of doing climate science that, again, is not new to science but has been gaining significance in the context of the epistemic challenges rooted in the increasing complexity in science: the *experience* of scientists in their field in general and especially in dealing with the particular instruments and methods (for example, the specific simulations they use).

This experience is what Lenhard (2018) describes as the "feeling" scientists have for the models, which can only be acquired in practice, in working in their area of expertise and in working with the models.

In a similar vein, one might interpret the emphasis philosophers and scientists place on understanding the history of the models and modelling practices. Computer simulations of this kind cannot be epistemically grounded fully in an ahistorical way. On the one hand, as Winsberg has pointed out, the credibility of the model is based in part on the techniques and practices of model construction through the tradition of their application. On the other hand, knowledge about the history of the models themselves can be crucial to circumventing, at least to some extent, the opacity the complexity of the models causes. Both cases highlight the relevance of the experience that the scientists have with handling the models in practice to increasing the credibility of models and modelling techniques.

Climate scientists themselves have also observed the significance of experience to their work (Tebaldi and Knutti, 2007). However, as we have seen in this chapter, in this context the notion of expertise, experience or expert judgement is also often associated with subjectivity. Though there is often a negative connotation to subjective decision-making, climate scientists also note its unavoidability.

But, when experience has such a significant relevance, then three questions have to be asked:

1. about the nature of this experience
2. how it is acquired
3. how it is justified.

These questions will be discussed in the next chapter. I will argue that a crucial element of this *experience* is tacit knowledge. A fundamental part in gaining expertise in any subject is grounded in experience, in having practiced in the

specific field in question and often learning from working in close proximity to others already having some expertise. It requires a knowledge that is more than what can be learnt from books and it is not easily put into words without any kind of 'showing' either. I will employ a broad interpretation of the term *tacit knowledge* that does not only apply to knowledge which cannot be expressed in principle but also the kind of knowledge that is not expressed in practice for whatever reasons. Tacit knowledge is not new to science, however, so far it has lived a mostly inconspicuous life in science. But, as I will argue in the next chapter, it gains greater significance in the context of highly complex systems where traditional approaches to knowledge acquisition are challenged.

The relevance of tacit knowledge also steers the focus onto the institutions that are at the centre of the scientific community because they are the place where this tacit knowledge is acquired, taught and communicated. *Institutions* has to be understood here in a broad sense, not just as specific organisations but as the extensive structures that make the scientific endeavour possible.

Winsberg (2018) has made a similar point by arguing for philosophy of climate science to turn its attention to the social structures of science. But he comes to this conclusion from a slightly different angle. Winsberg concludes, because of all of the epistemic problems and obstacles in climate science already discussed in this chapter,

> that philosophers do better to paint a picture in which we urge trust in the consensus of the scientific community, based on features of that community's social organization, then to try to provide a normative framework from which we can demonstrate the reliability (or its absence) of such-and-such modelling results. (Winsberg, 2018, p. 161)

By this Winsberg does not negate that it is, for instance, possible that "a simulation modeler could explain to his peers why it was legitimate and rational to use a certain approximation technique to solve a particular problem" (Goodwin, 2015, pp. 342–343; Winsberg, 2018, p. 161). But this kind of inner-scientific process of legitimising certain techniques, methods or hypotheses is always local and specifically context-dependent. Instead of trying to find schemes that would ground climate models normatively, philosophers, Winsberg argues, should rather focus on the social structures as "climate science is, in a thorough-going way, a socially organized kind of science, and [...] many features of its epistemology need to be surveyed at the social level in order to be properly understood" (Winsberg, 2018, pp. 209–210).

In the following I will also argue for turning the attention to the social organisation of science. For one, I agree with Winsberg that the social structure of science – specifically one that is as scattered between so many different individual scientists and institutions, dealing with highly complex systems and resulting epistemic problem as climate science – is significant to comprehending and grounding our trust in it. But I will argue as well that the increasing significance of tacit knowledge, or what Lenhard calls a "feeling" for the model, will also give new epistemic significance to the social structures in science.

4. Tacit knowledge, skill and expertise

When in conversation with outsiders to the climate-science community and asked how they go about making this or that decision, for example, about determining the possible range of a specific parameter in the tuning process, it is not uncommon for the climate scientists to explain that they "have experience" with the model they work with. This sort of experience forms a vital part, not just of climate science but of any kind of scientific endeavour.

It is a well-established insight in philosophy of science and epistemology that not all knowledge can be made explicit either for practical reasons or in more general terms. This knowledge, typically either called *tacit knowledge, non-propositional knowledge* or *knowing-how*, is considered to be an essential part of knowledge acquisition overall. However, considering that it is usually assumed not to be just part of everyday life but also crucial to science, the discussions about this 'phenomenon' in philosophy of science are relatively scarce. One explanation for this is that, although tacit knowledge is considered an indispensable feature of science, it is also an element of science that is "difficult to investigate" (Collins, 1974, p. 182). After all, tacit knowledge is often described as being the kind of knowledge that eludes explication for the person who is in possession of it. For instance, most people would say they "know how" to ride a bike when they are able to ride a bike down the street even though they might not be able to actually explain the *exact* physical principles making it possible for them to balance on a bike. Further, knowledge of those principles will not help the bike rider to be successful at riding a bike. Considering one can be in possession of knowledge that one at the same time (either in principle or for practical reason) cannot explain to someone else, it does not seem surprising that it might be challenging to make out exactly what constitutes this type of knowledge.

A second reason is that it often seems to be, in a way, uncomfortable for scientists and philosophers of science alike to admit that our understanding of

the world relies so significantly on a kind of knowledge that is hard to put into words. It goes against the self-perception of science, which is in many ways built on the notion that knowledge is independent from the specific scientists who acquires it. Conventionally, the replicability of scientific results (at least in theory) is considered to be essential to doing science (Fidler and Wilcox, 2021). However, case studies made by sociologist Harry Collins have shown it is not an uncommon occurrence in science that researchers can only recreate experiments successfully when they are directly demonstrated how to perform them. Thus, it might be more comfortable to focus on those aspects of knowledge that can be explicated, like it is done in scientific articles. Furthermore, for a philosophy of science that emphasises the context of justification, relying on knowledge that cannot be clarified or is at least difficult to clarify, is unsatisfying as it at most seems to be impenetrable for a full logical reconstruction of the argument.

Consequentially, a third reason is that the term *tacit knowledge* in itself is rather imprecise and not very well defined. The origin of the term is usually ascribed to Michael Polanyi ([1958] 1962, 1966a, 1966b). He sees tacit knowledge as an activity that is not just 'silent' but also one that 'cannot' be expressed. Since it was first developed, the term *tacit knowledge* has become common in other fields beyond philosophy, such as economics and management (Nonaka and Takeuchi, 1995). Besides *explicit* and *tacit knowledge* philosophers also often make use of the phrases *knowing that* and *knowing how* (Ryle, [1949] 1973). In a similar vein, the distinction of *propositional* and *non-propositional* knowledge is used.[1]

If one goes way back in the history of philosophy, some similarity can be found in the distinction between the concepts of *technê* and *epistêmê*. Fantl (2017) argues that at least the definition of the distinction found in Aristotle's *Nicomachean Ethics*, where *epistêmê* is usually translated as "scientific knowledge" and *technê* as "skill, art, or craft", can be seen as a predecessor to modern concepts of *knowledge how* and *knowledge that.*[2]

---

1 Other related concepts are the distinctions between *practical* and *theoretical* knowledge and *procedural* and *declarative* knowledge (for more information on the terminology, see Fantl, 2017).

2 Fantl (2017) especially sees parallels between Ryle's ([1949] 1973, p. 47) "knowing that" and "technê". He specifically refers to Aristotle seeing technê as "identical with the characteristic of producing under the guidance of true reason" (Nic. Eth. 1140a.10). Fantl concludes: "Such a conception of technê as skill guided by norms or rules anticipates

As far as modern philosophy is concerned, the concepts of *tacit knowledge* or *knowing how* are also often linked to Wittgenstein and Kuhn, who are considered to be "providing important insights into tacit knowledge and related epistemic issues" (Soler and Zwart, 2013, p. 7).[3]

After Polanyi and Ryle, the debate has (sporadically) been picked up by others in the fields of philosophy, history and social studies of science, most prominently by those coming from a background of the *new experimentalism* and the *practical turn*, whose representatives were most interested in questions of scientific practice. However, there have been few in recent years who turned specifically to the issue of tacit knowledge – one has to assume due to the problems already discussed above (see Soler, 2011).

Philosophy of climate science is here, with a few exemptions, no outlier. While, for instance, Winsberg (2018, p. 161) notes that there are some aspects of climate modelling that evade description, Lenhard (2020) mentions "the feeling" that climate scientists have for the models and Hillerbrand (2014, 2010) explicitly discusses non-propositional knowledge in climate-change uncertainty assessment. The significance of tacit knowledge in climate science has so far not been explored in more detail.

Climate scientists, on the other hand, point out on occasion aspects of their work that indicate an acknowledgement of these tacit components in the practice of climate science, even if they are not named so explicitly as will be discussed further below. This is unusual insofar as *tacit knowledge* has a bit of a bad reputation in science, at least as long as it comes to elements of justification procedures. The crux of the matter is that (at least in practice) tacit knowledge is usually difficult to make explicit and has a personal or subjective component; both features are conventionally not seen as signs of 'sound' science. Thus, even though tacit knowledge, as will be discussed below, is part and parcel of science, those aspects of science usually do not make it into scientific publications. However, in climate science the scientists themselves sometimes hint at those tacit features of their work. Therefore, one has to assume that the reliance

---

Ryle's identity of know-how with a disposition whose 'exercises are observances of rules or canons or the application of criteria' (Ryle, [1949] 1973, p. 47)" (2017).

3 Wittgenstein's (1953, 201) contribution is usually seen in his writings about rule following. Ludwig Fleck's ([1935] 1979) conception of "habits of thought" is also commonly seen as an early influence on the development of the idea of tacit knowledge.
Kuhn (1962, p. 44) himself refers to Polanyi in *The Structure of Scientific Revolution* arguing that the rules constituting a paradigm do not have to be made explicit in order for there to be a paradigm. See also Soler (2011, pp. 397–398).

on tacit knowledge is far more widespread. In fact, I will argue in the following that it is to be assumed that the necessity for tacit knowledge is significantly more prevalent in sciences that deal with additional epistemic challenges coming from highly complex systems. The claim I will make is that in those cases where the system explored and the instruments used are so complex that they are not fully transparent in all instances and all aspects to the scientists tacit knowledge gains an even more important role.

4.1 Tacit knowledge

In the following, I will briefly discuss the arguments made by Polanyi and Ryle as both are the most common reference points on the topic. Then I will also take a closer look at the in-detail analysis of tacit knowledge by sociologist of science Harry Collins, who explicitly discusses tacit knowledge in the context of modern science, before I will return to the topic of climate science and examine how specifically tacit knowledge applies there. Collins distinguishes three types of tacit knowledge, of which two, he argues, constitute tacit knowledge which could be made explicit at least in principle, but are not because either of the way society is structured or due to the limitations of our body. However, the goal is not to specifically explain in detail every single way that tacit knowledge is of significance in the context of climate science nor whether or not this tacit knowledge could, at least in principle, be made explicit. Rather it is to show how it permeates science at every step of the way and how the significance of this kind of knowledge increases under the framework of a science dealing with increasingly complex systems.

4.1.1 Michael Polanyi: tacit knowledge

Based on his personal experience as a chemist with a long and distinguished career, Michael Polanyi (1966a, p. 4) coined the term *tacit knowledge* to describe the circumstances that "*we can know more than we can tell*" (Polanyi, 1966a, p. 4). Polanyi's motivation is his opposition to an objectivist philosophy of science that sees science being a non-personal and non-subjective undertaking as a major characteristic of science (Polanyi, [1958] 1962, pp. 15–17). Contrary to the prevailing opinion of his time, Polanyi is convinced that knowledge in the end can only be understood as "personal knowledge" (Polanyi, [1958] 1962). For Polanyi, knowledge is personal insofar as it cannot be made fully explicit

and is based on experience and skill acquired in practice. He claims that "all knowledge is *either tacit* or *rooted in tacit knowledge*" (Polanyi, 1966b, p. 7).

There are two, now famous, examples from Polanyi's writings illustrating how he sees tacit knowledge operating and permeating every day life. The first example concerns face recognition:

> We know a person's face, and can recognize it among a thousand, indeed among a million. Yet we usually cannot tell how we recognize a face we know. So most of this knowledge cannot be put into words. (Polanyi, 1966a, p. 4)

Although we do not have the words to express how we recognise them, we, nevertheless, certainly possess a kind knowledge what the faces of people we are acquainted with look like and we put that knowledge to good use in daily life.[4] And there are ways to relay this knowledge, even though we have difficulty putting into words what makes us recognise a face. Polanyi specifically highlights the, at that time new, identikits used by the police to create pictures of suspects where witnesses can select from different templates of specific facial characteristics without having to give detailed descriptions of the suspect's face to an artist. However, our knowledge about other people's faces is not dependent upon the invention of techniques like this. The difficulty here is communicating the knowledge, not accessing it. Tacit knowledge is a kind of knowledge that one can be in possession of regardless of whether one finds a way to circumvent the linguistic barriers.

The second example concerns learning how to ride a bicycle. As already observed above, it is a common experience that one sometimes can do things, such as riding a bike, without needing to understand or be aware of the underlying (physical) processes:

> If I know how to ride a bicycle [...], this does not mean that I can tell how I manage to keep my balance on a bicycle [...]. I may not have the slightest idea of how I do this, or even an entirely wrong or grossly imperfect idea of it, and yet go on cycling [...] merrily. (Polanyi, 1966b, p. 4)

What is more, riding a bike (for a human being) can only be learnt by practicing it. One cannot learn how to do so by reading about it in a book. I can spend a lot of time studying the underlying physical principles making it possible for a

4 To understand how much this knowledge simplifies everyday interactions one only has to take a look at some accounts of people who suffer from face-blindness.

human being to balance on a bicycle, however, this will not give me the skill of being able to ride a bike (Polanyi, 1966b, p. 7).[5]

Yet Polanyi sees tacit knowledge not just as part of daily life but also part and parcel of science. Scientists, argues Polanyi, rely on the specific skill they developed in their specialist field. The acquisition of skill is a necessary and time-consuming part of the training as a scientist. Polanyi notes that skill is something that can be *"achieved by the observance of a set of rules which are not known as such to the person following them"* ([1958] 1962, p. 49). Experience is at the heart of this. And in science, like in other occupations requiring connoisseurship, it can only be obtained through practice and in company of those who already have the ability:

> To become an expert wine-taster, to acquire a knowledge of innumerable different blends of tea or to be trained as a medical diagnostician, you must go through a long course of experience under guidance of a master. Unless a doctor can recognise certain symptoms, e.g. the accentuation of the second sound of the pulmonary artery, there is no use in his reading the description of syndromes of which this symptom forms part. He must personally know that symptom and he can learn this only by repeatedly being given cases for auscultation in which the symptom is authoritatively known to be present, side by side with other cases in which it is authoritatively known to be absent, until he has fully realized the difference between them and can demonstrate his knowledge practically to satisfaction of an expert. (Polanyi, [1958] 1962, pp. 54–55)

Further, Polanyi also notes, and what will be significant further down below, that this also has implications for how the training of future scientists is done:

> The large amount of time spent by students of chemistry, biology and medicine in their practical courses shows how greatly these sciences rely on the transmission of skill and connoisseurship from master to apprentice. It offers an impressive demonstration of the extent to which the art of

---

5 Inspired by *Gestalt psychology*, Polanyi ([1958] 1962, pp. 53–55) sees tacit knowledge rooted in the distinction of *subsidiary awareness* and *focal awareness*. In the same way that a pianist has to concentrate on the entire piece of music they are playing (*subsidiary awareness*) and not on the specific actions their hands are performing (*focal awareness*) in order to successfully play music, tacit knowledge requires this kind of shift in focus awareness from the distinct to the whole, where attention must be unspecific and invisible so not to fail, Polanyi argues.

> knowing has remained unspecifiable at the very heart of science. (Polanyi, [1958] 1962, p. 55)

Thus, for Polanyi, in many ways tacit knowledge is central to practicing science. It is not just the primary way for an apprentice to acquire the necessary knowledge and skill that makes a scientist a scientist but also facilitates new scientific insight.

4.1.2 Gilbert Ryle: knowing how and knowing that

For the sake of completeness, it is worthwhile mentioning that, around the same time as Polanyi developed the idea of *tacit knowledge*, Gilbert Ryle came up with the related concept of *knowing how* and *knowing that* ([1949] 1973). Whereas Polanyi came to the issue from a philosophy-of-science perspective, Ryle looked at it from the point of view of philosophy of mind. What unites Ryle and Polanyi is an opposition to what they consider to be the dominant perspectives at that time in their respective fields. While Polanyi is concerned about an objectivist perspective on science, Ryle voices worry about the "intellectualist legend", which proclaims that "the intellectual execution of an operation must embody two processes, one of doing and another of theorizing" ([1949] 1973, p. 32). He claims that the intellectualist legend would ultimately lead into a vicious regress:

> The crucial objection to the intellectualist legend is this. The consideration of propositions is itself an operation the execution of which can be more or less intelligent, less or more stupid. But if, for any operation to be intelligently executed, a prior theoretical operation had first to be performed and performed intelligently, it would be a logical impossibility for anyone ever to break into the circle. (Ryle, [1949] 1973, p. 31)

Thus, Ryle argues that knowing how cannot, by default, require conscious reasoning as that would mean one would end in a situation where it is not clear how the first initial step should be initiated.

Instead, he sets out to offer a "positive account of knowing how" ([1949] 1973, p. 40). For Ryle knowing how to do something constitutes a disposition to behave a certain way:

> Knowing *how*, then, is a disposition, but not a single-track disposition like a reflex or a habit. Its exercises are observances of rules or canons or the applications of criteria, but they are not tandem operations of theoretically avow-

> ing maxims and then putting them into practice. Further, its exercises can be overt or covert, deeds performed or deeds imagined, words spoken aloud or words heard in one's head, pictures painted on canvas or pictures in the mind's eye. Or they can be amalgamations of the two. (Ryle, [1949] 1973, p. 46)

For Ryle, thus, like Polanyi *knowing how* is something that requires training or more general a practical learning process. Ryle also sees knowing how as an intelligent activity that is more than mere habit, instead it displays a degree of flexibility and adaptability to changes of circumstances (one might think of, for example, the car driver who reacts spontaneously in a perilous situation). This knowledge might have been obtained by some direct verbal instructions, but Ryle ([1949] 1973, pp. 47–50) emphasises that this does not mean that we do consciously follow these rules in our mind.

The debate concerning *knowledge how* and *knowledge that* and whether one can be reduced to the other is ongoing as an argument of *intellectualism* versus *antiintellectualism* in philosophy of mind (for an overview, see Fantl, 2017). In the following I will, however, be using the term *tacit knowledge*. Not least because it is the one most commonly used, not just by philosophers of science but also in science itself (insofar as it is discussed at all), while the dualism *knowing how* and *knowing that* is historically closer associated with debates in philosophy of mind. The term *tacit knowledge*, however, also conveys, in its opposition to the *explicit* or *explicable*, that it is a kind of knowledge that is, for practical or more fundamental reasons, not put into words, which will become an important feature in the case of (climate-)science practice discussed below. To that end, a closer look at specific aspects of the role of tacit knowledge in modern science seems prudent.

4.1.3 Harry Collins: a taxonomy of tacit knowledge

One person who has extensively explored the unique role that tacit knowledge plays in actual scientific practice in the recent decade is sociologist[6] Harry

6 As a sociologist, Collins sets a different goal for his analysis of tacit knowledge than a philosopher might do. Collins describes his approach as being "just a plumber" (2013, p. 26) in creating a scheme to explore and structure tacit knowledge. Collins (2010, p. 146) also specifically criticises most philosophical approaches to tacit knowledge for having put the human body at the centre of any investigation of knowledge.

Collins. In multiple case studies and over several decades, Collins (2014, 2013, 2001, 1974; Collins and Evans, 2009), specifically in the field of gravitational-wave physics, has studied how physicists rely on tacit knowledge in their everyday work life. Collins has also written broadly about the concept of *expert* and *expertise*, a topic that, as already discussed and will be further explored in the following, is intricately connected to tacit knowledge. Exploring what constitutes *expertise* also has specific bearings in the context of public perception of climate science, where the expertise of the scientists has been questioned in the past by those who wanted to sow doubt about anthropogenic climate change. In this context Collins provides a helpful framework to look at the intricate connection between expertise, practice, experience and tacit knowledge in the context of increasing complexity in science.

In his book *Tacit and Explicit Knowledge* (2010) Collins introduces a taxonomy of tacit knowledge that is useful to get an understanding of the variety of functions and forms that tacit knowledge can take in science. Collins broadly defines three different types of tacit knowledge, each referring to different intensities of 'tacitness' and a way in which something cannot be made explicit:

1. Relational Tacit Knowledge (RTK)
2. Somatic Tacit Knowledge (STK)
3. Collective Tacit Knowledge (CTK)

Before taking a closer look at each of these types of tacit knowledge, a few words need to be said about Collins' definition of tacit knowledge to avoid misunderstandings later. While for Polanyi the opposite of tacit knowledge is knowledge that is *explicable*, Collins defines *explicit* knowledge as the opposite to tacit knowledge. For Polanyi, tacit knowledge is that kind of knowledge that cannot be made explicit. Collins, on the other hand, defines tacit und explicit knowledge in the way it is transmitted:

> The tacit is communicated by "hanging around" with such persons. In children and older students tacit knowledge is acquired by socialization among parents, teachers, and peers. In the workplace it is acquired by "sitting by Nellie" or more organized apprenticeship. In science it is acquired during research degrees, by talk at conferences, by laboratory visits, and in the coffee bar. (Collins, 2010, p. 87)

That is, for Collins tacit knowledge is defined by being acquired through close proximity to those who already are in possession of this knowledge, whereas

explicit knowledge can be transmitted over longer distance.[7] However, this does not mean that explicit knowledge cannot also be transferred directly, in close contact, and that this might not enhance the learning process, for example, in a classroom situation, according to Collins (2010, p. 87). Further, he also recognises that some types of tacit knowledge could be transformed into explicit knowledge under the right conditions. As a matter of fact, from the three categories of tacit knowledge that Collins defines only the last one (*Collective Tacit Knowledge*) constitutes tacit knowledge that could not be turned into explicit knowledge, even in principle at some point in the future.[8]

4.1.3.1 Relational Tacit Knowledge

Relational Tacit Knowledge (RTK) is the weakest kind of tacit knowledge that Collins (2010, pp. 85–98) identifies. It refers to types of tacit knowledge that could theoretically be made explicit but is not done so in practice because of particular limitations of the structure of our society. It is knowledge that is

> experienced by humans as tacit knowledge and acquired as tacit knowledge, even though it is not the "ontology" of knowledge, nor even the structure of the human body and brain that have made them transferable in this way only. (Collins, 2010, p. 96)

---

7 Collins explains the transmission of explicit knowledge in terms of what he calls "strings". Strings are "bits of stuff inscribed with patterns: they might be bits of air with patterns of sound waves, or bits of paper with writing, or bits of the seashore with marks made by waves, or irregular clouds, or patterns of mould, or almost anything" (2010, p. 9). Though the strings themselves do not have meaning, they carry information that can be turned into meaning through interpretation of the strings. Collins argues that explicit knowledge is an economically "cheap" kind of knowledge because it can be "broadcasted" into the world at a considerable low cost (Collins, 2013, p. 27). However, Collins also point out that this does not mean that broadcasted explicit knowledge is automatically also understood. The "receivers of explicit knowledge have to be fluent in the language of the transmission medium and fluency in language is acquired as tacit knowledge" (Collins, 2013, p. 28). In this respect Collins agrees with Polanyi that all knowledge is tacit at its core.

8 Collins notes that in this context there are different meanings of the term *cannot*. He identifies eight different interpretations of "cannot" (2010, pp. 88–91). Some of these – that span from *logistic practice* and *technological impossibility* or *technical competence* to *somatic limit* and *contingency* – are of relevance in Collins' conceptions of tacit knowledge (see below).

This might happen for several reasons, argues Collins: sometimes knowledge is just kept concealed deliberately (*concealed knowledge*). For instance, it is not an uncommon occurrence that scientists from one lab try to conceal or at least not completely reveal their knowledge how to perform an experiment successfully. This knowledge could be put into words but is intentionally kept from others and, thus, could only be acquired by outsiders through "infiltrating the group" (Collins, 2010, p. 92). There is also that kind of RTK that is transmitted by directing the attention to a specific practice or object, for example, through touching or inspecting an object (*ostensive knowledge*). This knowledge could also be made explicit in theory but is too complex in practice. Further, there are situations where the logistics of the situations is so demanding that it is not feasible to turn it into explicit knowledge (*logistically demanding knowledge*). Such a situation might be the knowledge a worker in a big warehouse has who can locate every product in the warehouse in an instance by physically walking there, though they might have difficulty giving a description to someone else. Such a person could in principle be substituted by a computer system, but this might be considered to be too costly. In certain cases, knowledge is also kept tacit accidentally because there might be a misunderstanding concerning how much background knowledge the person who wants to acquire knowledge from another person has (*mismatched salience*). If person *A* tries to communicate *X* to person *B* and *A* assumes that *B* has some knowledge relating to *X* which *B* in fact does not have than *X* cannot be transmitted. Last but not least, Collins argues, there is that kind of RTK where a person *A* themselves is not certain how they actually perform a task insofar as they do not know what actions are actually important to succeed in carrying out the task, even though they are successful in doing it (*unrecognised knowledge*). Though the knowledge could in principle be made explicit, in this case, *A* is not able to do so because they are not aware of it. However, it is still possible that the relevant knowledge can be transferred through close proximity to *A* and even become explicit over time.

Even though RTK is neither in principle tacit nor will it in practice necessarily always remain so, Collins argues that one could still call it tacit as our experience of it is that it *is* tacit:

> In society as we know it there will always be secrets, mismatched saliences, and things that are unknown but may be about to become known. [...] the fact is that whatever you do there will always be knowledge that is not made explicit for these contingent reasons and it, therefore, will be an ever-present

> feature of the domain of knowledge as it is encountered even though its content is continually changing. (Collins, 2010, p. 98)

Collins notes that, though not all RTK could be made explicit at the same time, there is nothing preventing any individual piece of RTK to be made explicit in general. Therefore, according to Collins, the "principles to do with the nature of knowledge are not at stake" (2010, p. 98).

4.1.3.2 Somatic Tacit Knowledge

Somatic Tacit Knowledge (STK) is a stronger form of tacit knowledge than RTK (Collins, 2010, pp. 99–117). It is tacit knowledge that cannot be made explicit due to the limitations of the human body.

The most well known example for STK, according to Collins, is Polanyi's famous example of bicycle riding (see Chapter 4.1.1). Riding a bike is learnt through practice and usually through proximity to people who already know how to do so. And while it is possible to read and learn about the relevant physical laws in a book, this does not contribute to acquiring this particular skill. However, as Collins stresses, it is not impossible to imagine circumstances under which reading or being told about the physical principles of balancing on a bike might actually make it possible to acquire the skill to ride a bike:

> if our brains and any other elements of our physiology involved in balancing on a bike worked a million or so times faster, or, what is the equivalent, if we rode our bike on the surface of a small asteroid with almost zero gravity so everything happened much slower, we ourselves could probably use [...] rules to balance. Under these circumstances, balancing on a bike would be like assembling flat-pack furniture: as we began to fall to the left or the right we would consult a booklet and slowly adjust the angle of steering according to the instructions for remaining upright. (Collins, 2010, p. 100)

Abilities that rely on STK are usually carried out unconsciously and are often done better unconsciously, notes Collins (2010, p. 104). This might give an "appearance of mystery" (Collins, 2010, p. 117). But Collins claims that such concerns are unfounded.

For one, tasks that humans perform by relying on STK could still be done by artificial intelligence. For another, Collins notes that there are always things that specific objects or animals (including humans) are better at doing because

of the specific way they are built.[9] Thus for Collins humans rely on STK to perform certain complex tasks because of reasons that are inherent to *them as humans*, not the knowledge. He concludes that it would be a "mistake is to see all problems of human knowledge acquisition as problems of knowledge" (Collins, 2010, p. 105). STK just like RTK could, in principle, be made explicit, argues Collins, but is not done so due to the specific circumstances of being human (for example, having a limited brain capacity on this specific planet).

Another form of STK, Collin identifies, he demonstrates using the example of playing chess. While it is often claimed that computers can beat humans at chess, Collins argues that whether this is the case or not actually depends on how one defines *playing chess*, how one judges whether this task has actually been done by a computer. So far computers have only been able to beat humans at playing chess by a *brute force* approach. That means that the computer is able to calculate a few steep ahead of the humans through sheer computer power and some general heuristic, which is enough to win against the best human chess player. However, if one defines the ability of being good at playing chess not as "wining a game of chess" but as "playing the way humans play chess", then the answer to the question whether computers can beat humans at playing chess is a different one (Collins, 2010, pp. 106–113). Collins considers this to be the difference between what he calls "somatic-limit tacit knowledge" (winning a chess game) and "somatic-affordance tacit knowledge" (playing chess the human way). Humans, contrary to computers, rely on pattern recognition when playing chess. Until now computers have not been able to mimic this kind of pattern recognition, but, in theory, at least one could imagine a machine doing just that. What hinders us in creating such a machine at the moment is our inability to reproduce the functionality of a human mind or body.

Both kinds of STK can, thus, Collins stresses, at least in principle, be made explicit[10] but are not done for practical reasons. The reason that some researchers, nevertheless, consider this kind of tacit knowledge to be an ex-

9 As examples of this Collins notes that while humans are better at doing a lot of cognitive tasks such as calculating or copy-typing, than sieves, trees or dogs. Sieves commonly better sort stones and dogs are better at acting in reaction to smell than humans (2010, p. 105).

10 Collins defines *explicit* here as "expressed scientific understanding of causal sequences" (2010, p. 117).

ceptional kind lies for Collins in the importance we put on making things explicit:

> In sum, there is nothing philosophically profound about Somatic tacit knowledge, and its appearance of mystery is present only because of the tension of the tacit with the explicit: if we did not feel pulled towards trying to say what we do, and if we did not make the mistake of thinking this is central to the understanding of knowledge, we would find nothing strange about our brains' and bodies' abilities to do the things we call tacit. (Collins, 2010, p. 117)

Here, like in the case of RTK, the tacitness in STK is not insurmountable. However, the barrier to overcome might in practice be more challenging and it might not (yet) be possible.

4.1.3.3 Collective Tacit Knowledge

The third kind of tacit knowledge that Collins (2010, pp. 119–138) differentiates is *Collective Tacit Knowledge* (CTK). Contrary to RTK and STK, CTK is defined as a type of tacit knowledge that cannot be made explicit because it is solidly situated in the social sphere. It is the kind of tacit knowledge that is required not just to ride a bike but navigate it in traffic. Collins argues that it calls for a certain kind of knowledge to drive a car in traffic, where there are other drivers, that goes beyond knowing the traffic rules and knowing how to use a steering wheel or to change gears. Further, this kind of knowledge depends on where in the world one is. The experience of driving a car in China or Italy is quite different from driving in the UK and requires some "social judgment", notes Collins (2010, p. 122).

There is a certain "social sensitivity" and "degree of flexibility" (Collins, 2010, p. 123) needed for many things that we do on a daily basis. It is the thing we rely on when, for instance, we have to improvise. This type of knowledge is tacit in nature and, Collin argues, specific to humans insofar as we are able to interpret context-dependently:

> What is being argued is that humans differ from animals, trees, and sieves in having a unique capacity to absorb social rules from the surrounding society – rules that change from place to place, circumstance to circumstance, and time to time. (Collins, 2010, p. 124)

This knowledge is located in the realm of the collective social sphere, argues Collins. We all share in it, but we cannot possess it without being part of the

collective.[11] It is, according to Collins, an "enduring mystery" (2010, p. 123) how we have access to it. But as he concludes, it is a necessity to be human to take part in it, yet it is not essential to have a (full and able) human body. A person with a missing limb can still "know what it is to possess the collective human body shape [...] through the medium of a language that has been part formed through the physical interactions with the world of all those other human bodies" (Collins, 2010, p. 136).

Thus, one can also obtain CTK without actually participating in a collective practice, according to Collins. He calls this interactional expertise (see Chapter 4.2.1). That is, a sociologist of science could acquire interactional expertise about a subject just by being around and talking to scientists about how to do research in that particular field, even though the sociologist does not participate in that research. This means, Collins argues, that one can, at least in principle and after spending a significantly long time within the specific scientific community, engage in conversations on a highly specialised level without actually being scientists in that field.[12] In a similar way, leaders of big research project can acquire knowledge about various aspects of the project in order to make decisions about the research project's future without actually contributing any research. Though, Collins notes, it might sometimes still be helpful to engage in practice to acquire CTK, because of how our bodies or societies are constructed, it is "a matter of the nature of humans not the nature of knowledge" (2010, p. 138). But importantly, one still has to be immersed in the particular community.

---

11 Collins explicates this by a modified version of John Searle's (1980) *Chinese Room* thought experiment. The question that Collins puts forward is if it were possible for the person in the room to continue to engage in the exchange of questions and answers over a long period of time. Collins denies this because language is not fixed but dynamic and changes after a relatively short amount of time. This would make it impossible to pass as a native speaker to the people outside the room after a certain time.

12 An example Collins gives of such a situation from his personal life concerns how he has acquired interactional expertise in the field of gravitational wave research, which he has shadowed and observed for several decades as a sociologists. He claims to have actually managed to pass a kind of 'Turing Test' where he and an actual scientist separately and anonymously answered a number of in-depth questions concerning the research. The answers were then given to other experts in the field of gravitational physics who were not able to tell conclusively whether the answers were given by the actual gravitational-wave physicist or by Collins (Collins and Evans, 2009, pp. 104–109).

Compared to RTK and STK, CTK cannot be made explicit, even in principle, and there are no machines (we can imagine) that can imitate it, argues Collins:

> As far as knowledge is concerned, the deep mystery remains how to make explicable the way that individuals acquire collective tacit knowledge. We can describe the circumstances under which it is acquired, but we cannot describe or explain the mechanism nor build machines that can mimic it. Nor can we foresee how to built such machines in the way we can foresee how we might build machines to mimic somatic tacit knowledge. In the second case we know what we would need to do to make them work, in the first case we will not know how to start until we have solved the socialization problem. (Collins, 2010, p. 138)

For Collins CTK is the "central domain of tacit knowledge" (2010, p. 153).

4.2 Tacit knowledge in climate science

The reason for examining Collins' categorisation of tacit knowledge in detail here is that it illustrates nicely the variety of roles tacit knowledge can take, not because I now plan to move on to analyse every instance of tacit knowledge that might be significant in the working life of a climate scientists. In fact, I think this would be rather tedious and somewhat missing the point, considering that tacit knowledge by its nature is simultaneously omnipresent and frequently hard to detect.[13] In general, however, I agree with Collins' assess-

13 However, if one wants to better understand how tacit knowledge permeates all areas of science, in general it is worthwhile to first take a quick look at one of many case studies Collins did to explore tacit knowledge in the context of actual scientific practice. In this case study Collins (1974) examines the struggle of several different groups of physicists trying to construct a "Transversely Excited Atmospheric Pressure CO2 laser" (TEA laser) in the early 1970s. Collins observes the difficulties of a group of scientists to replicate a TEA laser just from reading the articles published on this subject by another, already successful research group. Only once the former got into contact with that later, through laboratory visits and other communication, do they figure out how to build a functioning TEA laser. The reasons for this, according to Collins, are manifold. For one, the scientists who originally created the laser were not so keen to outright reveal their knowledge due to competitiveness. But, as Collins emphasises, it also turned out (in hindsight) that the scientists had knowledge that they were not aware of initially but which was necessary to build the laser, which they were only able to pass along through showing others. Studying the publications on this topic was not merely

ment that all types of tacit knowledge he identifies are integral to doing science (2010, p. 150). Instead, the rest of this chapter is dedicated to the following two questions:

1. why the dependency of science on tacit knowledge is more visible in climate science than other more traditional fields of science
2. how and to what extent the pervasive presence of tacit knowledge can give us a definition of expertise that can function as workaround for the failed ideals from Chapter 3, as I implied at the end of Chapter 3.4

One particular feature of tacit knowledge that Collins' analysis has shown is how a lot of the knowledge tacit to us, or we acquire as such, might not be inherently tacit.[14] It is tacit for us because of some more mundane reason such as particular social structures or because of the limits of the human body to deal with significant complexity in an explicit way. Particularly the latter explains why the reliance on tacit knowledge is especially visible in climate science. It seems reasonable to assume that, when dealing with a system as complex as the climate system and equally complex models, scientists rely even more on tacit knowledge. The experience that scientist have with the models they work with or the "feeling", as Lenhard describes it, fulfils an important role, without which developing ESM would not be possible in practice. In such cases where, for instance, specific parameters are otherwise not very well constricted, the high complexity of the model makes it impossible to test the full range of possible parameter values as this would be far too time consuming. In these cases

enough to successfully recreate the TEA laser. In a later publication, Collins states that, although he had not yet developed the above classification at the time of the aforementioned case study, "building a TEA laser is a matter of RTK + STK + CTK" (2010, p. 152). See also Collins (2001) for a similar case study on tacit knowledge of the measuring of the quality factor of sapphire for the use in gravitational-wave detection.

14 I would like to note that while Collins might be right to claim that much of the knowledge that we come across as tacit is not tacit in principle but due to the limits of the human body or because of the way that society is (currently) structured. This might be right in principle. However, the assertion that knowledge might have an explicit form under quite different premises, such as on a different planet where people have a different brain capacity or in a completely differently structured society, might be useful when the aim is to point out that there is nothing 'mysterious' about this kind of tacit knowledge, as Collins (2010, p. 117) does. It is less so when one is concerned with science as it is done in practice at this point in time and the epistemological problems scientists are confronted with right now.

the *experience* with the models can be a helpful 'tool' scientists can resort to. More generally speaking, Alexander and Easterbrook conclude that climate-modelling institutions retain "a deep but tacit knowledge base about their own models" (2015, p. 1222; see also Easterbrook and Johns, 2009).

One might, nevertheless, come to the conclusion that these instances of tacit knowledge are only a feature of the process of the construction of models or development of experiments and question whether the insight that tacit knowledge is part of the daily practice of science has any implications on the justification of scientific research results. However, as already noted at the end of Chapter 3, in the context of climate modelling the realm of discovery and justification can no longer be separated as easily as such a suggestion might imply. Further and more significantly, as Léna Soler (2011) argues, tacit knowledge in general plays a significant role in the context of justification of procedures and products of science.[15] Soler emphasises that scientists develop a kind of "scientific 'sense' or 'instinct'" (2011, p. 406) that they make use of when, for instance, scientist O is faced with the question why they consider two experiments done at different times as 'the same' or why they decide at some point in the experimentation process there to be 'enough' evidence requiring no further testing:

> Faced with such questions, O will again, at some point, encounter insurmountable limitations in his attempts to clarify his reasoning. He will come to see that he is not able to put forward crystal-clear reasons. At some point, he will rely upon a personal intuition, a scientific 'sense' or 'instinct' which cannot be further analyzed by linguistic means and which refers to him as a particular individual. [...]
> O's intuition or scientific sense that is involved here can be viewed as a personal compass. This compass is not transparent, even to O himself. It points in a certain direction but it is a black-box (or at least contains some residual black boxes). We commonly assume that O's compass has been calibrated through O's previous experience, and that it has increased in sensitivity in proportion to the duration of O's first-person involvement in similar kinds of scientific practice. Moreover, we commonly assume that O's individual,

15 Soler defines justification in this context in the following broad way: "'S provides justifications in favor of X' means: 'S gives his own motives to believe X or to perform X'" (2011, p. 407).

> specific talents might play a role. However, the process of regulating the compass remains largely opaque. (Soler, 2011, p. 406)

It is easy to see similarities between the "compass" described here that scientists draw on when assessing the merits of an experiment and the "feeling" that Lenhard describes climate researchers establish for the models they work with.

Soler argues that all of this leads to an "opacity of experimental practice" (2011, p. 403) that goes beyond an opacity in the realm of discovery and has to be seen as contrary to the widespread "rationalist ideal of completely self-transparent knowledge" demanding "a fully explainable justification of human knowledge, a justification in which no step would be left in the shadows, in which each link in the reasoning chain could be exhibited and scrutinized" (2011, pp. 406–407). This opacity is at least in actual scientific practice, if not more deeply, anchored at the core of science insofar as it concerns a kind of scientific 'intuition' – though a consequential part of experiment development and justification – that is rarely attempted to discuss or make more explicit (Soler, 2011, p. 413).

One specific place where this kind of experience plays a particularly visible role in the context of climate science is the reliance on expert judgement to assess different lines of evidence. That is, evaluating the strength and weaknesses of different data sets and types of data, different types of models and ensembles or methods (such as emergent constraints) as discussed in Chapter 3.3.3.4). This requires, as Zickfeld et al., note not just assessing the specific literature but also "knowledge that is not explicit in the formal literature" (2007, p. 237). Common expert judgement when evaluating MIPs, for example, concerns assumptions about the quality and independency of different models (Hillerbrand, 2010; Lee et al., 2021, p. 568).

It seems reasonable to assume that this is a kind of background information that is primarily acquired in practice, not just for pragmatic reasons, but also because it requires some knowledge that is at least difficult to make explicit as it is a very context specific synthesis of a wide variety of pieces of information.

Lam and Majszak come to a similar conclusion in an analysis of the role of expert elicitation in the identification of tipping points (critical thresholds, which once crossed result in considerable, oftentimes abrupt and irreversible change to the climate system) about the necessary expert judgements in this process:

> in many cases it seems related to the experts' own experience and interpretation of certain nonclear-cut, possibly ambiguous, situations. For instance, this knowledge may involve practical experience of model behavior, interpreting ambiguous data and the relative relevance of feedback processes, drawing connections and building links between disciplines, among other things. (Lam and Majszak, 2022, p. 8)

Climate-change assessment is more than a simple calculation. Instead the scientist's expertise developed over time through their experience of working with the models and creating data sets is a significant and non-neglectable aspect to evaluating models and observations as well as assessing climate-change hypotheses.

Because expert judgement is usually seen as something 'subjective', concerns have been raised in the respect to how the elicitation of expert judgements is handled and structured expert elicitation protocols have been proposed with the aim to avoid or mitigate this 'subjectivity' (Hanea et al., 2021; Oppenheimer et al., 2016; Thompson et al., 2016). While there are certainly advantages to such procedures as making the selection of experts more explicit and possibly reducing some specific biases[16], it seems questionable such a protocol could ever make expert judgements fully transparent, as these expert judgements in themselves are still fundamentally based on the tacit knowledge gained through the practical experience of the scientists.[17]

To be able to judge the adequacy of a scientific argument, more is needed than just reading journal articles. Having specific tacit knowledge is constitutive to being a scientist. However, this also gives us the option to draw a connection between tacit knowledge and a concept of *expertise*.

16 Lam and Majszak (2022), however, note that, considering the variety of ways social values can play a role in the model building and evaluation processes, such structured expert elicitation would make it not a value-free process (see Chapter 3.1.3) and that there are also certain value-laden trade-offs to be made in the development of these protocols.

17 The argument I have made here for the most part concerns tacit knowledge in the process of climate-model development and evaluation. However, one must assume that tacit knowledge, experience and skill take a similarly prominent role in the gathering and evaluation of observational data in the same way that skill has been noted to be an important quality of a successful experimenter. Anecdotally, I can report that in a conversation with a young climate scientist talking about her work, when asked how she would go about filtering noise from the actual signal, she answered if she did not know, she would "ask older, more experienced" scientists at the institute.

1.2.1 Connection between tacit knowledge and expertise

Collins (2014), Collins and Evans (2009), and Collins, Evans and Weinel (2016) distinguish two kinds of expertise that characterise scientists (and other professionals and people that have acquired a distinct skill).[18] A closer look at these two types of expertise, *contributory expertise* and *interactional expertise* will be helpful to understand how expertise is intricately connected to tacit knowledge. It will also shed some light on the question we were left with at the end of chapter 3 of what actually constitutes an expert.

Both are forms of *expertise* which require *specialist tacit knowledge*[19] but show, according to Collins and his co-authors, differences in the way they can be accessed and utilised. Contributory expertise refers, as the name already says, to those who provide a piece of knowledge to an area of specialist expertise "and is, generally, what people think of when they hear the word 'expert'" (Collins, 2014, p. 64). Collins emphasises that this kind of specialist expertise requires practice. One becomes a contributory expert by becoming an apprentice and by practicing in the specific field of expertise, in the company of others who are already experts in this field and learning from their abilities. As Collins puts it: "one does not become 'a scientist' without practice, and a lot of practice" (2014, p. 58).

The immersion in the scientific community cannot be substituted by reading scientific books and journal articles. Although one might (theoretically) ac-

---

18 From this perspective, being a scientist requires no different type of expertise than that which, for instance, a doctor or an engineer has. But this kind of expertise can also be attributed, e.g., patients with rare chronic diseases who not uncommonly develop "knowledge about the treatment of those diseases that compares with or even exceeds that of their doctors" (Collins, 2014, p. 64).

19 The authors also acknowledge that there are other more ubiquitous kinds of expertise including "all the endless indescribable skills it takes to live in a human society" (Collins and Evans, 2009, p. 16), that is, skilful abilities we all have but which are often not considered to be noteworthy. Further, Collins and Evans argue there are also kinds of specialist expertise that are solely build on *ubiquitous tacit knowledge*, not *specialist tacit knowledge*, such as popular understanding of science or knowledge acquired by reading scientific papers without being a member of the scientific community. These kinds of expertise, however, have clear limits, as discussed in this chapter. Collins and Evans also point out a meta-expertise that enables discrimination between two or more experts (2009, pp. 18–23). The problems, particular for laypersons to recognise expertise are discussed further below.

cumulate substantial knowledge[20] this way, as Collins and Evans note, it also bears a significant risk of misjudging the material at hand:

> what is found in the literature, if read by someone with no contact with the core-groups of scientists who actually carry out the research in disputed areas, can give a false impression of the content of the science as well as the level of certainty. Many of the papers in the professional literature are never read, so if one wants to gain something even approximating to a rough version of agreed scientific knowledge from published sources one has first to know what to read and what not to read; this requires social contact with the expert community. Reading the professional literature is a long way from understanding a scientific dispute. (Collins and Evans, 2009, p. 22)

This can also cause problems for effective science communication, when some research is of particular public interest and people who have no specific training in the particular field of research but, nevertheless, consider themselves experts because they have read some papers and are convinced they can judge the adequacy of the reasoning process behind the argument without any training as a specialist or current inclusion in the specific scientific community. Some of the most prominent climate-science critics have been scientists who also claim to be experts in the connection between smoking and cancer, the origins of acid rain and the increase of the ozone hole (Oreskes and Conway, 2010). Considering what it takes to become a true expert in these times, it is doubtful that they have actually acquired specialist expertise based on specialist tacit knowledge in the way described here in all those quite different research topics.[21] Assuming Collins' and Evans' assessment of the connection between tacit knowledge and expertise is right, it seems prudent to assume that in these cases these climate science critics are, amongst other things[22], actually missing the required tacit knowledge vital to assessing reasoning processes in

20 Collins and Evans call this kind of knowledge *primary source knowledge* (2009, pp. 22–23).

21 Oreskes and Conway write about the scientists in question (most prominently Fred Singer and Fred Seitz), though once "prominent researchers" in their own rights, "had no particular expertise in environmental or health questions" and "did almost no original scientific research on any of the issues" they attacked (2010, p. 8).

22 Oreskes and Conway (2010) also uncover not just strong financial ties between this group of scientists and specific interest groups from the affected industries but also strategies to artificially amplifying their voices. They further ascribe the scientists a strong personally motivated rejection of any kind of governmental regulation.

science which can only be acquired by being immersed in a specific scientific community. In a similar vein, it is necessary to be part of the scientific community to know which people working in and around the field in question are considered to be serious experts and what reputation the specific journal a scientific paper is published in has. All of this is crucial knowledge to judge the adequacy of an argument that cannot be simply gained from reading papers.

Hence, just reading books and papers clearly does not make one a specialist. However, this might make one question whether it does not significantly limit the number of people who can judge the adequacy of scientific arguments. Here the second kind of specialist expertise which Collins and Evans define, *interactional expertise*, comes to into play. This term refers to "the expertise in the *language* of a specialism in the absence of expertise in its *practice*" (2009, p. 28). Absence of practice, however, does not mean that interactional expertise can be acquired in isolation. It still requires immersion into the specialist community to obtain the necessary tacit knowledge and is, thereby, far from being a quick and easy undertaking.

One instance from the history of climate science where not all people involved were in possession of the required interactional expertise, also highlighted by Collins (2014, pp. 80–91) was the *Climategate* 'scandal' in 2009 (see Chapter 1). Interactional expertise, argues Collins, was needed to know that the "trick" the scientist were talking about in the leaked emails was not an attempt to mislead the public about the severity of climate change through deliberately and illicitly manipulating data. According to Collins, one needs 'inside information' about the 'language' that climate scientists speak to know that *trick* had a different meaning than the common connotation of 'deceiving'. This is something that one can only learn when associating with the specific community of scientists, not from reading some journal articles.

While interactional expertise can be acquired all on its own without engaging in practice, like for instance a sociologist of science who spends years with scientists of a specialised field, this is rather the exception, Collins and Evans note (2009, pp. 104–109). The much more common way to acquire international expertise is through establishing contributory expertise, argues Collins. In science interactional and contributory expertise are usually obtained together as "learning to become a contributory expert in a narrow technical domain is mostly a matter of acquiring interactional expertise because it is through talk that one learns how to act in practical matters" (2014, p. 72). In fact, Collins argues that interactional expertise fulfils a highly important role

in science and "is key to most of what happens in science" (Collins, 2014, p. 72). Interactional expertise is, for instance, what makes it possible for scientists to evaluate the arguments made by other scientists in peer-review processes, without having done exactly the same research (Collins, 2014, p. 72; Collins and Evans, 2009, p. 60), although the interactional expertise referred to here is established most likely in the process of acquiring contributory expertise.

Interactional expertise gains particular significance in times where the increasing complexity in research subjects and questions means a widespread distribution of the workload between different researchers and research groups. In modern scientific research projects, specifically those requiring a high number of scientists working on one and the same problem, scientists can never be contributory experts in every aspect but still have to be able to communicate with the other scientists in the project. Collins discusses this using the example of gravitational wave physics:

> There are around a thousand physicists working in the international, billion-dollar field of gravitational-wave detection. Each of them belongs to a sub-specialism within the area, [...]. In the main, no person from one subgroup could step in and do the work of a person from another subgroup – at least not within a long apprenticeship. If that were not so, they would not be specialists. And yet all these people have to coordinate their work. The way they coordinate their work is by sharing a common language which they learn when they attend the many international conferences that are part of their job and by visiting and spending time at each other's laboratories. What they are doing is acquiring interactional expertise in each other's specialities. (Collins, 2014, pp. 69–71)

It is easy to see how this relates to climate science. Climate simulations are commonly a product of many hundreds of scientists' contributions over more than one generation. Institutions that develop climate models are, usually subdivided into many different working groups, specialising in different aspects of modelling the atmosphere, ocean and land and so forth.

To coordinate this work, it requires regular meetings between the heads of different working groups. Especially considering the interdependency of different model components (see Chapter 2.1), so no research group for a particular model component can do their work in isolation from the other ones; coordination and organisation are key. Improvements and changes in the different model components done willy-nilly could set the whole model array. Interactional expertise provides scientists with a "common language" to negoti-

ate theses issues. Similarly, interactional expertise also makes discussions and cooperation with scientists from adjacent areas of science possible.

Another element related to this kind of management and internal communication work in science is the expertise developed in other research projects or/and in other research field and then 'transferred' to the current conundrum, which Collins and Evans (2009, pp. 64–66) call *referred expertise*. This kind of meta-expertise enables the scientist, specifically those in leading positions, to judge how to proceed in a (large) research project:

> The experience in other fields is applied in a number of ways. For example, in other sciences they have worked in, they will have seen that what enthusiasts insist are incontrovertible techniques turn out to be controvertible; this means they know how much to discount technical arguments. [...] They will have a sense of how long to allow an argument to go on and when to draw it to a close because nothing new will be learned by further delay. They will have a sense of when a technical decision is important and when it is not worth arguing about. They will have a sense of when a problem is merely a matter of better engineering and when it is fundamental. (Collins and Evans, 2009, p. 66)

Thereby, *referred expertise* is a kind of expertise that goes beyond but is also facilitated through *interactional expertise*; it is, however, different to *contributory expertise*, which is always distinctly local.[23]

Expertise, or at least the expertise we are interested in here, requires social engagement. This stands in contrast to the common ideal (or maybe more accurately the 'caricature') of the lonely, unsocial scientists working on his own in a lab. An ideal that is not very close to what is actually going on in science. Scientists work in community because the questions posed by modern science are just too complicated to be solved by just one person, but also because it is the place where expertise is gained, established and refined. "[W]hile something can be learned from instruction books and other kinds of literature, the heart of an expertise is acquired by picking up tacit knowledge", thus, by being in company of those who are already in possession of it (Collins, 2014, p. 60).

23 Collins and Evans note, that managing scientific research projects, of course, also requires all sorts of non-science specific expertise in terms of "financial management, human resources management, networking skills, political skills, and so forth; some of these will comprise the contributory expertise of management itself" (2009, p. 66).

Tacit knowledge has always been an essential part of science. This might not have been acknowledged as much in the past because the "rationalist ideal of completely self-transparent knowledge", as Soler (2011, p. 406) calls it, is strong in science and epistemology alike and the role of tacit knowledge was easier to overlook, specifically to people outside the scientific community. However, in the context of sciences dealing with more and more complex questions and systems and the resulting added epistemic difficulties, the significance of tacit knowledge also becomes more and more visible. In climate science this can be seen not just in the tacit knowledge needed in communicating and organising research in its widely dispersed state but also in the tacit knowledge coming into play when climate scientists exert expert judgement and in the "feeling" scientists develop for models. The complexity of the climate system and the models moves the dependency of science on tacit knowledge further into the 'visible spectrum'.

Thinking of expertise and tacit knowledge in this manner can be a helpful way out of the dilemma we were left with at the end of Chapter 3, where it became apparent that certain ideals about how science ought to operate that are usually appealed to as a guarantee for 'good science' fall short in the context of increased complexity in modern science. As tacit knowledge is at the centre of many of the methods and practices that are in contradiction to the aforementioned ideals, reconceptualising tacit knowledge not as something lacking a kind of transparency science requires but as fundamental to all knowledge and the basis of any kind of scientific expertise can instead ground these practices. I will come back to this in a bit, but first, I want to take a detour to look at one specific way the increasing relevance of tacit knowledge manifests and becomes visible in the case of climate science.

4.2.2 Climate modelling as engineering or craft

One way to examine the increasing specific relevance of tacit knowledge in climate science is to look at some of the descriptions climate scientists themselves use for their work. What becomes noticeable very quickly is that scientists (and philosophers of science for that matter) often revert to words that characterise climate modelling as something akin to an engineering task and/or requiring some kind of creativity. The most striking example of this is the tuning of models where the process has been repeatedly compared to a "craft" or an "art" (e.g., Mauritsen et al., 2012; Hourdin et al., 2017; see also

Edwards, 1999). In one of the most well-known papers about tuning Mauritsen et al. write:

> The model tuning process at our institute is artisanal in character, in that both the adjustment of parameters at each tuning iteration and the evaluation of the resulting candidate models are done by hand, as is done at most other modeling centers. (Mauritsen et al., 2012, p. 16)

Though the terminology is not uncontroversial. In an article called "The art and science of climate model tuning", Hourdin et al. (2017) state that despite the title that there is some ambivalence among the authors whether *art* is the appropriate term to describe the process of tuning:

> There was a debate among authors on the idea of using the word art in the title of the paper. Tuning is seen by some modelers more as a pure engineering calibration exercise, which consists of applying objective or automatic tools based on purely scientific considerations. Others see it as an experienced craftsmanship or as an art: "a skill that is attained by study, practice, or observation." As in art, there is also some diversity and subjectivity in the tuning process because of the complexity of the climate system and because of the choices made among the equally possible representations of the system. (Hourdin et al., 2017, p. 598)

Nevertheless, Hourdin et al. also link tuning to what is commonly considered a distinct artistic practice, i.e., being a conductor of an orchestra:

> Climate model tuning is a complex process that presents analogy with reaching harmony in music. Producing a good symphony or rock concert requires first a good composition and good musicians who work individually on their score. Then, when playing together, instruments must be tuned, which is a well-defined adjustment of wave frequencies that can be done with the help of electronic devices. But the orchestra harmony is reached also by adjusting to a common tempo as well as by subjective combinations of instruments, volume levels, or musicians' interpretations, which will depend on the intention of the conductor or musicians. (Hourdin et al., 2017, p. 590)

The comparison of climate-model tuning to reaching harmony is also interesting insofar as music is often considered to be an endeavour that requires some kind of tacit knowledge in the learning process (Polanyi, [1958] 1962, p. 56). For instance, it is hard to imagine how one should be able to learn how to play a trumpet without ever having hold a trumpet, just by reading or listening to

instructions. Many of the characteristics of what is commonly attributed to a good musician, specifically one who makes music as part of a group, fall within the realm of what Collins calls CTK. It requires creativity, intuition and often the ability to improvise; all skills that are characterised by eluding explicability and also being fundamentally human in its nature.

As we have seen in Chapter 3.4.3 scientists concerned about tuning note that any kind of procedure that renders tuning in more automatic terms will not be able to fully rule out the subjective, artisanal aspect of tuning. It merely moves the subjective decision making to a different level. Scientists still have to make judgement calls concerning trade-offs (Mauritsen et al., 2012, p. 16). The subjective and personal expertise of the scientists thus is an unavoidable component of climate modelling.

But comparing techniques of model building to an art or a craft also draws attention to another aspect of climate modelling. A craft is something that has to be learnt through apprenticeship and requires training as well as experience. A successful craftsman is someone who has acquired expertise in a skill through exercising that skill. It emphasises that tuning complex computer simulations calls for being well acquainted with the model in question. Experience in working with the model is vital (Hourdin et al., 2017, p. 398).

This fits with a more general description of climate modelling overall "without any pejorative connotations intended whatsoever, as engineering, or even tinkering" (Held, 2005, p. 1611). This comparison seems valid not just in respect to tuning but more generally considering that many of the epistemic issues of climate modelling arise from features of software engineering such as modularity and kludging, as discussed in Chapter 2 and 3. Similarly the process of developing parametrisations can be described as "akin to an engineering problem" (Parker, 2018), in respect to the task of finding a way of adequately implementing a process that cannot be integrated into the model in a resolved way. As these models are not fully theoretical constructs, some aspects of climate modelling also have elements of a trial-and-error approach such as the iterative method of model development and evaluation. Particularly the latter has also been noted to have a creative element to it (Guillemot, 2010).

Struggling to fit computer simulation into traditional schemes of theory on the one side and experiments on the other side, William Goodwin points out that, while climate-science modelling does not adhere to these dualistic structure, climate science does resemble applied sciences and engineering:

> many of the same issues that arise in thinking about how it is possible to make reliable predictions about out future climate also arise when trying to understand how engineers are able to make reliable estimates of the flight characteristics of wings that no one has ever built, or to calculate the effects of turbulence in the pipes of a proposed chemical plant. (Goodwin, 2015, p. 346)

However, it should be noted that, contrary to what Goodwin implies, climate models are not just employed to assess anthropogenic climate change but are also used to explore much more fundamental question about the climate system in the same way that 'traditional' sciences do (Parker, 2018). There are also other differences to typical applied and engineering sciences, such as the degree to which both disciplines "apply techniques in ways that might turn out to be outside of the domain under which they have being directly tested" and the applicability as well as reliance on a *V&V* approach (Winsberg, 2018, p. 162, see Chapter 3.2.3.3). So the comparison to applied sciences has its limits and one should maybe resort to a more careful wording and say that climate-model developing involves some methods, techniques and epistemic obstacles resembling those known from applied sciences. However, what has been shown here is that the comparison of computer simulation development to engineering is especially used when describing that there is an element of 'trial and error', 'tinkering', 'skill', 'craft' or even outright 'tacit knowledge' as it is also commonly associated with questions of engineering or technology (see also Franssen et al., 2018).

4.3 Conclusion: expertise through experience

Philosophers of science interested in complex computer simulations have long noted the "epistemic opacity" and lack of "analytical understanding" that comes along with these kinds of simulations (Humphrey, 2004; Lenhard and Winsberg, 2010). These philosophers are mostly concerned with the prospect of acquiring understanding (or the lack thereof) in respect to the internal processes of the models and/or the relationship between model and target system. However, there is also a different kind of opacity that is not so new to science and does not just turn into an issue for science where science hits a "complexity barrier" (Lenhard, 2019) that can only be circumvented through computer modelling. Soler (2011) argues that there is an inherent

opacity to any experimental practice in the sense that it is not possible for the experimenter to make the reasoning-process behind whether or not to accept an experiment as successful fully explicit (Soler, 2011, p. 404). Instead scientists develop through practicing their skill as a scientist what is described as a "compass" (Soler, 2011) or "feeling" (Lenhard, 2020), which functions as a substitute for explicable knowledge in these situations. This makes tacit knowledge an unavoidable and necessary feature of science. Though computer simulations are not comparable to traditional experiments in all respects (Winsberg, 2003), the opacity Soler alludes to here effects, as we have seen, the practice of climate modelling as well and is only intensified by the complexity of the models. Tacit knowledge has always been part and parcel of science – as all of human life taking place in community – but its presence becomes much more visible once a reduction in analytical understanding due to high complexity comes into the picture. The opacity of experimental or modelling practice, however, does not mean that this prevents scientists from assessing the work of their colleagues. Quite to the contrary, as Collins and Evans (2009) have argued, without expertise rooted in specialist tacit knowledge it is not even possible to evaluate the research of others, e.g., in peer-review processes. In these kinds of situations again the "feeling" (Lenhard, 2020) or "compass" (Soler, 2011) that scientists develop by participating and being part of the specific scientific community play a non-neglectable role.[24]

Despite the essential role that tacit knowledge takes in scientific practice, tacit knowledge is often an uncomfortable topic for scientists, specifically in those instances where science is under constant scrutiny from the public. The idea that every decision, every reasoning process can be made explicit, so it can be assessed by anybody, is deeply ingrained into how scientists see their work.

Nevertheless, as long as scientists are among themselves, the significance of tacit knowledge might not stand out very much. After all, all involved are in possession of the necessary tacit knowledge or, where it is missing, it can be acquired (by lab visits, for instance). However, once outsiders (or sometimes insiders) to the specific scientific community start voicing doubt, the impossibility to make everything that is going on in science explicit becomes apparent. The image of science as fully transparent and logically traceable to the last corner shows cracks.

24 This is not to say that often disputes among scientists can arise because of lack of some specific kind of tacit knowledge (Soler, 2011). However, these disputes are settled by (amongst other things) relying on tacit knowledge.

The point I would like to stress here is that the reliance on specialist tacit knowledge can be interpreted as a strength of science, much more than a weakness. It is, in the end, something that can be learnt. In that sense there is nothing 'mysterious' or 'esoteric' about it. It is not an ability that 'falls from the heaven' or that is only bestowed upon a few chosen ones. But to acquire the necessary skill to become an expert, one needs to acquire tacit knowledge which requires time and effort and being immersed in the specific specialist community, which (at the very least theoretically) anyone could access.

Tacit knowledge is part of everyday life. It is, to paraphrase Polanyi, at least at the root of all knowledge (Polanyi, 1966b). It is, for instance, what makes it possible for us to use language and take part in conversations. What sets the knowledge of experts, e.g., scientists apart from the tacit knowledge of daily life is that it requires *specialist* tacit knowledge (Collins and Evans, 2009, p. 14). One reason we rely so heavily on the expertise of others in all areas of life is that we cannot, for reasons of time constraints, obtain the necessary tacit knowledge in every instance.[25] What scientists, thereby, accumulate in their professional life is exactly that – together with more explicable knowledge – through training and immersion in the specific scientific community. This is the way that scientists commonly acquire expertise.

Before turning to the question what defining expertise in this way means for public controversies about science, there are two things with respect to the role of tacit knowledge in science that I would like to point out here.

First of all, recognising that tacit knowledge is fundamental to science is not to say that explicit or explicable knowledge does not also have a prominent and significant place in science. Getting some new piece of explicit knowledge is usually the ultimate aim of a research project. Particularly considering that, as has been noted in Chapter 3.2 and 3.3, many climate scientists conclude that improving explicit mechanistic understanding both of the climate system and the models is a way forward to further secure knowledge of future climate change. Still, while explicit knowledge is what science thrives towards, achieving it is in practice only possible through tacit knowledge.

Further, tacit knowledge, while being an indispensable feature of science is not what makes science *science*. What characterises science in general or more specific scientific disciplines are particular methodologies, rules, conventions,

25 In our daily life we, e.g., take advantage of the specialist tacit knowledge of others when we trust doctors to interpret ultrasound or X-ray scans.

etc.[26] Instead of a claim about the definition of science, the point I would like to make here is that tacit knowledge facilitates access to these methodologies, rules and conventions. That is, without the required specialist tacit knowledge one cannot acquire the necessary expertise to do science.

In some public debates in the last years, the use of the term *expert* has become almost derogative. The claim that the people are not in need of experts, that 'ordinary people' know better than experts – who do not seem to know anything anyway because they all seem to change their opinion all the time or because there seems to be no consensus even about critical, basic questions – or even worse that experts are all 'in cahoots' in order to suppress 'common folk' has been a reliable by-product of many public debates about scientific research. Particularly when these discussions are also connected to debates about policies which are perceived to be freedom-constricting and costly.

There are many reasons that such arguments enjoy a certain popularity in certain circles. One contributing factor, for sure, is the discrepancy between the representation of specific scientific debates in the media, where controversies and lack of consensus are artificially inflated (see Chapter 1).[27] A common problem that observers of the public climate-change debate have noted, for instance, is that for a long time the issue was often reported in the same way political arguments are conveyed: by purporting objectivity through reporting on both sides of an argument equally, negating that facts do not come with many sides (Oreskes and Conway, 2010, p. 7). Similarly, it is often more attractive for journalist to report on controversies than stable consensus.[28]

But another factor to consider contributing to the rejection of expertise, I would argue, is that *expert* is generally not a very well defined term. This makes

26 I will not define these rules and methodologies here any further, because, as Chapter 3 has shown, they show a certain adaptability and are always unique to a scientific community at a specific point in time as research objectives and questions often change over time. And, as we have seen in this chapter, whether or not these rules, methodologies and conventions are observed can (in the end) only be evaluated by members of said specific scientific community.

27 There are of course also particular psychological factor that make particular groups of people especially susceptible to reject experts, such as that accepting and following the expert advice would mean having to restrict one's personal life in a way that would be perceived as inconvenient and uncomfortable (see Chapter 1).

28 As has also been noted by journalists themselves see Rusbridger (2015).

determining whom to trust exactly as an expert more difficult for laypeople. It simultaneously leads to the problem of non-experts being in a position to claim the title, and at the same time 'ordinary people' not knowing which attributes to look out for to identify potential experts, particularly when it seems like there are many conflicting positions.

This leads us to the question: how does one then as a layperson recognise an expert? After all, the only fail-safe way to judge the expertise of others is by becoming an expert yourself. But are there ways from an external perspective to discriminate experts from non-experts? Having noted the connection between expertise and tacit knowledge, Collins and Evans (2009, p. 68) propose to see specialist expertise as directly connected to *experience*. Defining a specialist expert in this way, they argue, has the advantage, compared to other prominent criteria for judging expertise based on credentials[29] or track record (e.g., Goldman, 2001), that it does not exclude those instances where people acquire expertise without being formally trained, e.g., in the form of an university degree.[30]

However, I would argue, emphasising the experience as the distinctive feature of expertise has further benefits. This definition acknowledges that having acquired specialist tacit knowledge is the foundation of any (scientific) expertise. Putting experience and skill front and centre brings science practice and the institutions facilitating it into focus. It underlines the importance of being trained in something, being part of the scientific community and the social structures underneath all of this for becoming a scientific expert.

And most importantly it gives an (at least partial) answer to the question we were left with at the end of Chapter 3: how can a layperson discriminate between conflicting expert opinions. Chapter 3 has shown that neither specific methods nor virtues scientists bring to the job are an adequate way to determine what constitutes 'good' science. This chapter, on the other hand, has

29 Although it should be noted that experience and credentials, of course, often coincide, see also Chapter 5.

30 A prominent example of such a case is the expertise gained by activists during the AIDS-epidemic in the 1980s, Collins and Evans argue. These activists managed to acquire expertise that was on par with that of scientists working on potential treatments, in order to get into a position to advocate for quicker access to a possible effective drug. They were even able to contribute to the research, due to their unique knowledge about the habits of the patients (Collins and Evans, 2009, pp. 52–53; Epstein, 1995). For other examples see: Collins and Evans (2009) and Collins (2014).

highlighted the relevance of tacit knowledge to doing science. Resorting to *experience* as a criterion of expertise can be seen as the logical conclusion. When assessing whom to trust, determining expertise defined in this way is also significantly easier (though, of course, not infallible) for an outsider to the scientific community, while it is almost impossible for a layperson to assess if any internal standards or methods are adequately followed. The analysis of tacit knowledge in science done in this chapter has also shown that it cannot be the job of the public to pass judgement on the quality of the specific work done by scientists; the experience and knowledge to do so lie with the scientists themselves and it would be presumptuous to assume that a layperson could, on the spur of the moment, acquire the knowledge scientist need many years to amass in order to evaluate a scientific argument.

A potential counterargument against defining expertise in connection with experience, which I like to get out of the way here, is the claim that, as Kuhn (1962) has prominently argued in the history of science, scientific progress was often brought on by younger scientists. First, the young scientists might not have that much research practice, but they still have commonly gone through some sort of apprenticeship program. In science this usually means attending university, acquiring several degrees and by doing so becoming a member of and practicing in this community but there are also other ways. Secondly, the new perspective younger scientists bring to the debate is what makes them disagree with more established scientists. This perspective also constitutes a kind of experience. Thus, it is more a question of different forms of access to experience.

Nevertheless, expertise as experience, of course, is not a fail-safe way for laypersons to assess whom to trust; an expert or (more probable) a group of experts with a lot of experience can, of course, be wrong. In fact, this happens all of the time. After all, it is a hallmark of good science that it revisits knowledge in light of new evidence. But in the absence of any other criteria, it gives a good (first) indication. However, if one changes the question slightly and does not ask how to recognise an expert but how to recognise someone who claims to be an experts but actually is not, experience is a much more promising criterion. That is, it provides a good, practical strategy to 'sieve out' specific types of apparent experts that actually do not have any experience working in the field in question and/or are not immersed in the specific scientific community. As already discussed above, many prominent critics of climate science who were

'sold' to the public as experts by so inclined stakeholders are actually lacking this specific experience of working in climate science.

Expertise characterised in this manner, puts specialist tacit knowledge in the form of skill at the centre and makes it fundamental to doing science. This has advantages and disadvantages. The advantage is that it is a very inclusive definition, as it also includes those people who did gain expertise through unconventional channels. It further provides a good guideline when to be sceptical of claimed expertise. The disadvantage is that it does not provide a foolproof method for laypeople to identify experts. In my opinion it seems however highly questionable if it would ever be possible to establish a procedure or mechanism that would allow us to do so. As has been shown in this and the last chapter, the subjects and methods of science are just too complex to make this very likely. In the end, who is an expert and who is not can best be determined from within science. Only there the necessary tacit knowledge is given to make such judgements.

2. Concluding remarks

In the last four chapters, I have argued that certain ideals about how science operates cannot be fulfilled by modern science. It has been shown, by example of climate science, that this becomes particularly apparent in the context of science that deals with very complex systems. These ideals are, nevertheless, very pervasive in public debates about the reliability of certain scientific research. This makes it easy for specific interest groups that would like to avoid regulations to take advantage of the widespread presence of these ideals to sow doubt about particular research results. So fostering these ideals in the public understanding of science is in the interest of these stakeholders, which is one reason why these ideals are so persistent in the public perception of (climate) science, despite the fact that science can rarely if ever live up to them.

While these ideals might take slightly different forms in different public debates about different fields of science, three popular ones, in the form in which they arise in the context of the public debates about climate research that were examined here, concern: the value-freeness of science, the relationship between theory and model, and observations and the ability of science to produce clear, binary predictions.

In the process of investigating why these ideals cannot be applied to climate science, many of the epistemological difficulties of climate science were discussed. However, it should be again emphasised: the notion that there are particular epistemic challenges does not mean that these are so overwhelming that they prevent climate scientists from making meaningful and useful statement about the climate in general or anthropogenic climate change more specifically. Instead, what has been shown in this book is that the process of evaluating and estimating climate-change hypotheses and making projections and predictions is not an impossible task, but it is more complicate than often portrayed. One example of such a hypothesis that was discussed here concerns the value of ECS. It was shown how – with careful reasoning and taking into

account many different lines of evidence – scientists can come to epistemically well-justified conclusions despite all epistemic hurdles. In this context it was also pointed out that improvements in understanding the inner-workings of the models and target system is seen by climate scientists and philosophers as essential to increasing confidence in climate change related hypotheses (Bony et al., 2013; Knutti, 2018; Winsberg, 2018).

I have also argued that one aspect of scientific practice, in the context of scientists dealing with highly complex systems, is the increasing relevance but also visibility of specialist tacit knowledge. As Chapter 4 has shown, while climate scientists, on the one hand, have to deal with the epistemic opacity of the climate models and their relationship to the target system, the scientists can (at least to a certain degree) circumvent these obstacles through the "feeling" or "compass" that scientists acquire by working with their models (Lenhard, 2020; Soler, 2011). Scientists acquire this skill through their training, sustain it by practicing as scientists and expand it in conversation with and in company of other scientists. What the investigation of the role of tacit knowledge has shown is that the quality and reliability of scientific research results can only adequately be assessed by other scientists who work in the same or an adjacent field because only they are in command of the necessary specialist tacit knowledge.

However, I have also argued that recognising the role of tacit knowledge in science also gives us access to a practically useful criterion for when outsiders to a specific scientific community have good reason to be sceptical about a claim made by an apparent 'expert'. That is, when they make a claim that is contradictory to something the majority of other experts agree on and they further have no experience of practicing in the specific research field. In my opinion connecting expertise and experience in this way offers a helpful framework for how to think about expertise considering that the above described ideals cannot be consulted to determine whether or not to trust the scientists in question. Some practical implications of this will be discussed in the following.

5.1 Where to go from here?

By way of concluding, I will examine what follows from the failure of the three ideals about science and the increasing significance of tacit knowledge in more general terms. In order to do so, I will look at the situation from three different perspectives:

1. philosophy of science
2. science
3. public understanding of science

5.1.1 Philosophy of science

Firstly, considering the role of philosophy of science, I agree with Winsberg's assessment that philosophers will not "be able to tell a normatively grounded story that will secure the unassailable reliability of the results of climate modelling" (Winsberg, 2018, p. 161). There is no overall argument philosophers can make to the effect that climate models provide reliable and trustworthy results. Whether the models are reliable is a question wrongly put. Instead, the question needs to be whether the models can assess or contribute to assessing a specific hypothesis; that is, whether the models are adequate-for-purpose (Parker, 2009; Winsberg, 2018, p. 202). Thus, any concrete epistemological deliberation about the reliability of climate models is only possible on a distinctly local level and only with the caveat that such elements as expert knowledge and methodologically not fully constrained decision making are not fully accessible to the philosopher. Nevertheless, philosophers can contribute to getting a better understanding of epistemological difficulties in climate science and how they arise. However, climate scientists still remain "the best experts regarding what *should be believed*" (Winsberg, 2018, p. 163). That is, philosophers might reconstruct certain reasoning structures as we have seen in Winsberg's case study about RA and ECS. However, determining whether the necessary conditions for accepting a specific hypothesis is fulfilled is a task that can only be accomplished by scientists who have the relevant expertise.

For this reason Winsberg has advocated for philosophy of science instead to concentrate on the underlying social structures when trying to foster trust in climate science (2018, p. 161). I come to a similar conclusion, albeit from a slightly different angle. In Chapter 4 I have argued that tacit knowledge is not just at the root of all knowledge but that its relevance as well as its visibility increases the more complexity becomes an epistemic obstacle. Thus taking a closer look at the social structures and institutions, which facilitate the acquisition of this tacit knowledge, can be a helpful way to 'circumvent' the opacity the tacit-knowledge component of science leaves behind, as the social structures and institutions of science usually perform an important role in safeguarding against inadequate science. Better understanding how these structures and in-

stitution function can also foster trust in the scientific process, especially because, as I will argue below, they further understanding what specific features and characteristics can help identify experience in an expert.

Some specific aspects of the social structure of climate science have already been scrutinised by philosophers of science. One specific institution that has attracted quite a bit of attention from philosophers of science in this context is the IPCC. One reason for this is the unique position of the IPCC as a scientific institution, which provides policymakers with reports, summaries and reviews of the current state of research (the IPCC does not do its own research) but to some degree also integrates policymakers into the assessment process. This is quite unusual for science, which usually tries to draw quite a distinct 'demarcation line' between itself and politics, so as to not give the impression to be value-laden. Therefore, philosophers of science have shown specific interest in those aspects of the IPCC where the sphere of politics and of science intersect, such as the review process, where governments are invited to participate (Kosolosky, 2015) and the rules by which the authors are chosen (Leuschner, 2012b).

Some other aspects that are noteworthy in climate science from a social epistemology perspective that philosophers of science have turned their attention to concern, how to establish that there is actual consensus among experts (see Intemann, 2017) and what the epistemic consequences of distributed epistemic labour mean for attributing authorship when no single author can possibly be in possession of the whole range of necessary knowledge (Huebner et al., 2017). The question what influence public scrutiny has on how climate scientists do their work and communicate their knowledge to the public has also gained the attention of philosophers of science. In respect to the problem of artificially manufactured doubt in public climate-change debates (Oreskes and Conway, 2010), Biddle and Leuschner (2015), for instance, examine when dissent has an epistemic beneficial effect and when not.[1] As stated above, these kinds of research topics cannot just improve our understanding of the role of tacit knowledge but can contribute to learning how to attribute expertise to those who actually are experts.

1 See also Winsberg (2018, pp. 208–226) for a good first overview on different debates concerning climate science from the perspective of social epistemology.

5.1.2 Science

Secondly, it seems clear that science does not do itself any favours by upholding obsolete ideals instead of openly communicating how actual scientific research is done. In the short term it might seem convenient to resort to one or more of these ideal in order to strengthen one's argument. Insisting on the specific virtue of either the scientists or the scientific methods might be an easy way out of a particular debate. The ideals represented in Chapter 3 are attractive to science; they turn science into something that is 'above' other human undertaking, seemingly irrevocable and 'objective'. However, in the long run, these ideals can easily – as has already happened, not just in the case of climate science – be turned against science, by those wishing to undermine specific research because it has ethical, social or political implications that would be inconvenient to those agents.

More open communication from all branches of sciences about the actual process would make it harder to attack one specific field of research as an outlier. If science as a whole refrains from resorting to these ideals in science communication and instead choses to explain thought processes, background assumptions, methods, procedures and uncertainties more openly, it would make it harder to sell the 'failure' of some scientists to follow these ideals as a distinct misconduct of those scientists.

What makes this difficult is that these kind of ideals are also widespread among scientists. Specifically, the ideal of science being a value-free enterprise is a well-established assumption in science. Asserting the opposite, that is, that science cannot avoid value-decisions is often refuted vocally (see, e.g., Schmidt and Sherwood, 2015). Here research being *value-laden* is generally mixed up with the research being *biased* (Winsberg, 2018, pp. 150–151). Overcoming this wrong preconception is a difficult task as it is deeply rooted both in science and in the public understanding of science. On the other hand, as has been shown in Chapter 3.1, methodological not fully constrained decisions become more ubiquitous and tracing all decision-making processes, to lay open the reasoning process behind them, becomes impossible in practice with increasing complexity. Thus, it will become inevitable that science will be more vulnerable to attacks of apparent inappropriate value-ladenness, the more complicated the issues sciences tries to tackle become, specifically when the research has significant social relevance.

The other two ideals are less deeply ingrained in science. The notion that observations are underdetermined and that the relationship between model

and observational data is complicated is not new to science, as has been shown in Chapter 3.2. On the other hand, the relationship between theory and observation is often oversimplified in situations where scientific research results are communicated to the general public. Here a concept of observations, akin to a "direct empiricists" (Lloyd, 2012) perspective on science, is often draw upon to emphasise the trustworthiness of a research result. Accompanying problems like underdetermination, theory-ladenness and measurement uncertainties are rarely talked about in this context. Here, again, oversimplifying the relationship between observation and theory risks making science vulnerable to attacks from outside forces.

As far as expressing of uncertainties goes, climate science has done a lot over the years to become more adept at dealing with and communicating uncertainties though there are still some difficulties. The calibrated language as stated in the IPCC's *Guidance Note for Lead Authors* is the best example for this. Finding a consistent but also easily understandable framework for how to communicate uncertainties has been a long and not yet finished project (see also Landström, 2017). As has been noted in the most recent IPCC report, while a clearly calibrated language is employed, the terminology adopted in the Guidance Note still leaves room for misinterpretations (Chen et al., 2021, pp. 168–171).

Another aspect concerning uncertainty, where some scientists as well as philosophers see room for improvement that also infringes on the topic of communicating uncertainties in the IPCC report, concerns *expert judgement*. The climate science community has widely acknowledged the relevance of expert elicitation in making uncertainty assessments. There are also many proposals how to make the process of expert elicitation more explicit. However as I have argued in Chapter 4 the question remains if particular schemes of structured expert elicitations would not also risk just shifting at least some background assumptions to another level and ignore the fundamental tacitness of these kinds of judgements.

It has been a well-documented tactic from climate change sceptics to call for 'better' science and emphasising uncertainties in order to argue that it is still to early for regulatory policy. This is in stark contrast to the well-established insight shared both by philosophers of science as well as scientists that science is always fallible and that in that respect scientific research results are always preliminary. Further, as Howe (2014) points out scientists giving in to these demands has proven counterproductive in the past, hindering progress on taking actions. This does not mean that climate science should not try to

proceed to reduce uncertainties, rather the history of climate change policy has shown how risky giving in to these excessive and unsatisfiable public expectations can be.

5.1.3 Public

And thirdly, on the flipside of this coin, when it comes to the issue of public understanding of science, it has become clear that the lack of insight into actual scientific practice makes it easy for specific interest groups to sow doubt about scientific research results disadvantageous to them. One way of counterbalancing this is on the level of science education – to not just teach scientific knowledge but also knowledge about the methods, procedures and structures of science. Improving public understanding of these feature of science would further the trust in science. Here philosophy of science could also play an active role in facilitating the exchange between science and the public.

The second aspect regarding the general public's relation to science discussed in this book concerns expertise and how to recognise it. I have argued for a definition, introduced by Collins and Evans (2009), that defines expertise through experience in a specialist field. This definition at least gives those who are not members of a specific scientific community a criterion when to be sceptical about the claim of 'apparent' experts, specifically when their claim is contradictory to that of the majority within the scientific community. It offers a pragmatic solution to the gap left by the failure of the ideals discussed in Chapter 3. In Chapter 4 I have noted how in the context of climate science prominent climate sceptics who declare themselves 'experts' in the field of climate science actually have not practiced in the field of climate science. This is not a problem that is unique to public debates of climate research but also a feature of other public disputes about science.[2]

In Chapter 4 I have stated, following the argument by Collins and Evans (2009), that other conceptions of expertise like track records or reputation can

---

2 Oreskes and Conway's book "Merchant of Doubt" also discusses other such cases from the second half of the 20th century beyond the specific case of climate science. For a more recent example one might also look at the COVID-19 pandemic. During the pandemic, e.g., the German virologist Christian Drosten has publicly voiced concern about the way that (otherwise not named) scientist from other disciplines were passed off as experts in the media, despite them only having knowledge of the topic in question (that is, coronaviruses) "that does not go above superficial textbook knowledge" (Henning and Drosten, 2020).

potentially exclude certain kinds of people who actually have expert knowledge. Nevertheless, reputation or track records can be good indicators of experience. After all, as already discussed, the main way that this expertise in science is acquired is through studying the subject at university, by practicing in the field and being a member of the scientific community. This usually goes along with gaining a reputation, e.g., through specific career steps or through authoring publications. Publications can also be a way to gain some knowledge of the track records of scientists.[3] However, one has to keep in mind that reputation as well as track records on their own can be misleading as a criterion of when to consider someone an expert. For one, there are, as noted in Chapter 4, cases where people who have not gone through the traditional 'channels' of specialist expertise acquisition (e.g., getting an university degree) but, nevertheless, have assembled specialist expertise through other routes. For another, it might also not always be possible for non-experts to correctly evaluate the track record or reputation, because it requires, for example, knowledge which journals are taken seriously by the specific scientific community and which are considered fringe journals. Here an outsider to the scientific community might risk wrongfully assuming expertise.[4] That is, the criterion of experience is not a fail save principle by which a layperson can 'separate the wheat from the chaff', however, it can be a case-specific pragmatic solution to get some idea when to be sceptical about the claimed expertise of someone. Further, an additional assessment of the specific social structures of climate science can be helpful to better understanding how such characteristics as track records and reputation come about. In this sense an analysis of the social structures of climate science cannot just help to further understanding of the acquisition and relevance of tacit knowledge and experience in science but also play a role in strengthening trust in climate science.

---

3 It also has to be noted that there are also other features that are more external to the scientific process, which can be an indicator for inadequate research, such as when the research is financed by stakeholders who have an interest in a specific research outcome. However, as scientific research is increasingly financed by industry, how science is financed is also not a failsafe way to evaluate the reliability of the research and can, from an outsider's perspective, merely give an indication when there is disagreement amongst scientists (for examples of such cases of inadequate research financed by special interest groups in respect the climate science, see Oreskes and Conway (2010), for an example from medical studies, see Douglas (2000)).

4 For examples of such cases, see Collins (2014).

References

Alexander, Kaitlin, Easterbrook, Steve M., 2015. The software architecture of climate models: a graphical comparison of CMIP5 and EMICAR5 configurations, in: *Geoscientific Model Development* 8, 1221–1232. https://doi.org/10.5194/gmd-8-1221-2015

Arabatzis, Theodore, 2006. On the Inextricability of the Context of Discovery and the Context of Justification in: Schickore, J., Steinle, F. (Eds.), *Revisiting Discovery and Justification. Historical and Philosophical Perspectives on the Context Distinction*. Dordrecht: Springer, pp. 215–230. https://doi.org/10.1007/1-4020-4251-5_13

Aristotle, 1962. *Nicomachean Ethics*, Oswald, M. (Trans.). New York: Bobbs-Merrill.

Arrhenius, Svante, 1896. On the influence of carbonic acid in the air upon the temperature of the ground, in: *The London, Edinburgh, and Dublin Philosophical Magazine and Journal of Science* 41, 237–276. https://doi.org/10.1080/14786449608620846

Bacon, Francis, [1620] 1863. The New Organon, in: Spedding, J., Ellis, R.L., Heath, D.D. (Trans.), *The Works VIII*. Boston: Taggard and Thompson.

Baumberger, Christoph, Knutti, Reto, Hirsch Hadorn, Gertrude, 2017. Building confidence in climate model projections: an analysis of inferences from fit: Building confidence in climate model projections, in: *Wiley Interdisciplinary Reviews: Climate Change* 8, e454. https://doi.org/10.1002/wcc.454

BBC, 2018. Trump: Climate change scientists have "political agenda". 15 October 2018, https://www.bbc.com/news/world-us-canada-45859325 (accessed 17 May 2023).

Betz, Gregor, 2013. In defence of the value free ideal, in: *European Journal for Philosophy of Science* 3, 207–220. https://doi.org/10.1007/s13194-012-0062-x

Biddle, Justin B., Leuschner, Anna, 2015. Climate skepticism and the manufacture of doubt: can dissent in science be epistemically detrimental?, in:

European Journal for Philosophy of Science 5, 261–278. https://doi.org/10.1007/s13194-014-0101-x

Bjerknes, Vilhelm, [1904] 2009. The problem of weather prediction, considered from the viewpoints of mechanics and physics, Volken, E., Brönnimann, S. (Eds. And Trans.) in: *Meteorologische Zeitschrift* 18, 663–667. https://doi.org/10.1127/0941-2948/2009/416

Boé, Julien, 2018. Interdependency in Multimodel Climate Projections: Component Replication and Result Similarity, in: *Geophysical Research Letters* 45, 2771–2779. https://doi.org/10.1002/2017GL076829

Böhm, Reinhard, Auer, Ingeborg, Brunetti, Michele, Maugeri, Maurizio, Nanni, Teresa, Schöner, Wolfgang, 2001. Regional Temperature Variability in the European Alps: 1760–1998 from Homogenized Instrumental Time Series: Regional Temperature Variability in the European Alps, in: *International Journal of Climatology* 21, 1779–1801. https://doi.org/10.1002/joc.689

Bony, Sandrine, Stevens, Bjorn, Held, Isaac, Mitchell, John F., Dufresne, Jean-Louis, Emanuel, Kerry A., Friedlingstein, Pierre, Griffies, Stephen, Senior, Catherine, 2013. Carbon Dioxide and Climate: Perspectives on a Scientific Assessment, in: Asrar, G.R., Hurrell, J.W. (Eds.), *Climate Science for Serving Society*. Dordrecht: Springer, pp. 391–413. https://doi.org/10.1007/978-94-007-6692-1_14

Boykoff, Maxwell T., Boykoff, Jules M., 2004. Balance as bias: global warming and the US prestige press, in: *Global Environmental Change* 14, 125–136. https://doi.org/10.1016/j.gloenvcha.2003.10.001

Brient, Florent, 2020. Reducing Uncertainties in Climate Projections with Emergent Constraints: Concepts, Examples and Prospects, in: *Advances in Atmospheric Sciences* 37, 1–15. https://doi.org/10.1007/s00376-019-9140-8

Briffa, Keith Raphael, Schweingruber, Fritz H., Jones, Philip Douglas, Osborn, Timothy John, Harris, Ian C., Shiyatov, Stepan Griorevich, Vaganov, Eugene A., Grudd, Hakan, 1998. Trees tell of past climates: but are they speaking less clearly today?, in: *Philosophical Transactions of the Royal Society of London. Series B: Biological Sciences* 353, 65–73. https://doi.org/10.1098/rstb.1998.0191

Brönnimann, Stefan, Wintzer, Jeannine, 2018. Climate data empathy, in: *WIREs Climate Change* 10, e559. https://doi.org/10.1002/wcc.559

Brysse, Keynyn, Oreskes, Naomi, O'Reilly, Jessica, Oppenheimer, Michael, 2013. Climate change prediction: Erring on the side of least drama?, in: *Global Environmental Change* 23, 327–337. https://doi.org/10.1016/j.gloenvcha.2012.10.008

Caldwell, Peter M., Zelinka, Mark D., Klein, Stephen A., 2018. Evaluating Emergent Constraints on Equilibrium Climate Sensitivity, in: *Journal of Climate* 31, 3921–3942. https://doi.org/10.1175/JCLI-D-17-0631.1

Carrier, Martin, 2013. Values and Objectivity in Science: Value-Ladenness, Pluralism and the Epistemic Attitude, in: *Science & Education* 22, 2547–2568. https://doi.org/10.1007/s11191-012-9481-5

Carrier, Martin, 2010. Scientific Knowledge and Scientific Expertise: Epistemic and Social Conditions of Their Trustworthiness, in: *Analyse & Kritik* 32, 195–212.

Cartwright, Nancy, 1983. *How the Laws of Physics Lie*, Oxford: Oxford University Press. https://doi.org/10.1093/0198247044.001.0001

Charney, Jule G., Arakawa, Akio, Baker, D. James, Bolin, Bert, Dickinson, Robert E., Goody, Richard M., Leith, Cecil E., Stommel, Henry M., Wunsch, Carl I., 1979. Carbon Dioxide and Climate: A Scientific Assessment, in: *National Academy of Science* 22.

Chen, Deliang, Rojas, Maisa, Samset, Bjørn H., Cobb, Kim, Diongue-Niang, Aida, Edwards, Paul, Emori, Seita, Faria, Sergio Henrique, Hawkins, Ed, Hope, Pandora, Huybrechts, Philippe, Meinshausen, Malte, Mustafa, Sawsan Khair Elsied Abdel Rahim, Plattner, Gian-Kasper, Treguier, Anne Marie, 2021. Framing, Context, and Methods, in: Masson-Delmotte, V., Zhai, P., Pirani, A., Connors, S.L., Péan, C., Berger, S., Caud, N., Chen, Y., Goldfarb, L., Gomis, M.I., Huang, M., Leitzell, K., Lonnoy, E., Matthews, J.B.R., Maycock, T.K., Waterfield, T., Yelekçi, O., Yu, R., Zhou, B. (Eds.), *Climate Change 2021: The Physical Science Basis. Contribution of Working Group I to the Sixth Assessment Report of the Intergovernmental Panel on Climate Change*. Cambridge, New York: Cambridge University Press, pp. 147–286.

Churchman, C. West, 1948. Statistics, Pragmatics, Induction, in: *Philosophy of Science* 15, 249–268. https://doi.org/10.1086/286991

Clark, Andy, 1987. The kludge in the machine, in: *Mind and Language* 2, 277–300.

Collins, Harry, Evans, Robert, Weinel, Martin, 2016. Expertise revisited, Part II: Contributory expertise, in: *Studies in History and Philosophy of Science Part A* 56, 103–110. https://doi.org/10.1016/j.shpsa.2015.07.003

Collins, Harry M., 2017. *Gravity's Kiss. The Detection of Gravitational Waves.* Cambridge: MIT Press.

Collins, Harry M., 2014. *Are we all scientific experts now?* Cambridge: Polity Press.

Collins, Harry M., 2013. Building an Antenna for Tacit Knowledge, in: *Philosophia Scientiae* 17, 25–39. https://doi.org/10.4000/philosophiascientiae.882

Collins, Harry M., 2010. *Tacit and Explicit Knowledge*. Chicago; London: University of Chicago Press.

Collins, Harry M., 2001. Tacit Knowledge, Trust and the Q of Sapphire, in: *Social Studies of Science* 31, 71–85.

Collins, Harry M., 1974. The TEA Set: Tacit Knowledge and Scientific Networks, in: *Science Studies* 4, 165–185.

Collins, Harry M., Evans, Robert, 2009. *Rethinking Expertise*. Chicago; London: University of Chicago Press.

Collins, Matthew, Knutti, Reto, Arblaster, Julie, Dufresne, Jean-Louis, Fiechefet, Thierry, Friedlingstein, Pierre, Xuejie, Gao, Gutowski Jr., William J., Johns, Timothy C., Krinner, Gerhard, Shongwo, Mxolisi, Tebaldi, Claudia, Weaver, Andrew J., Wehner, Michael, 2013. Long-term Climate Change: Projections, Commitments and Irreversibility, in: Stocker, T.F., Qin, D., Tignor, M., Allen, S.K., Boschung, J., Nauels, A., Xia, Y., Bex, V., Midgley, P.M. (Eds.), *Climate Change 2013: The Physical Science Basis. Contribution of Working Group I to the Fifth Assessment Report of the Intergovernmental Panel on Climate Change*. Cambridge: Cambridge University Press, pp. 1029–1136.

Cook, John, Nuccitelli, Dana, Green, Sarah A., Richardson, Mark, Winkler, Bärbel, Painting, Rob, Way, Robert, Jacobs, Peter, Skuce, Andrew, 2013. Quantifying the consensus on anthropogenic global warming in the scientific literature, in: *Environmental Research Letters* 8.

Daston, Lorraine, 2011. The Empire of Observation, 1600–1800, in: Daston, L., Lunbeck, E. (Eds.), *Histories of Scientific Observation*. Chicago: University of Chicago Press, pp. 81–113.

Daston, Lorraine, Galison, Peter, 2007. *Objectivity*. New York: Zone Books.

Daston, Lorraine, Lunbeck, Elizabeth, 2011. Introduction: Observation Observed, in: Daston, L., Lunbeck, E. (Eds.), *Histories of Scientific Observation*. Chicago: University of Chicago Press, pp. 1–9.

Dee, Dick, Uppala, Sakari M., Simmons, Adrian J., Berrisford, Paul, Poli, Paul, Kobayashi, Shinya, Andrae, Ulf, Balmaseda, Magdalena, Balsamo, Gianpaolo, Bauer, Peter, Bechtold, Peter, Beljaars, Antonius C. M., van de Berg, Leo, Bidlot, Jean, Bormann, Niels, Delsol, Claire, Dragani, Rossana, Fuentes, Manuel, Vitart, Frederic, Geer, Alan J., Haimberger, Leopold, Healy, Sean, Hersbach, Hans, Hólm, Elías Valur, Isaken, Lars, Kallberg, Per, Köhler, Martin, Matricardi, Marco, McNally, Anthony P., Monge-Sanz, Beatriz M., Morcrette, Jean-Jacques, Park, B. Kevin, Peubey, Carole, de Rosnay, Patricia, Tavolato, Christina, Thépaut, Jean-Noël, Vitart, Frederic,

2011. The ERA-Interim reanalysis: Configuration and performance of the data assimilation system, in: *Quarterly Journal of the Royal Meteorological Society* 137, 553–597. https://doi.org/10.1002/qj.828

Douglas, Heather, 2009. *Science, Policy, and the Value-Free Ideal*. Pittsburgh: University of Pittsburgh Press.

Douglas, Heather, 2004. The Irreducible Complexity of Objectivity, in: *Synthese* 138, 453–473. https://doi.org/10.1023/B:SYNT.0000016451.18182.91

Douglas, Heather, 2000. Inductive Risk and Values in Science, in: *Philosophy of Science* 67, 559–579.

Duhem, Pierre, 1906. *La théorie physique: son objet, sa structure*. Paris: Chevalier & Riviére.

Dunlap, Riley E., Jacques, Peter J., 2013. Climate Change Denial Books and Conservative Think Tanks: Exploring the Connection, in: *American Behavioral Scientist* 57, 699–731. https://doi.org/10.1177/0002764213477096

Easterbrook, Steve M., Johns, Timothy C., 2009. Engineering the Software for Understanding Climate Change, in: *Computing in Science & Engineering* 11, 65–74. https://doi.org/10.1109/MCSE.2009.193

Edwards, Paul N., 2010. *A Vast Machine. Computer Models, Climate Data, and the Politics of Global Warming*. Cambridge: MIT Press.

Edwards, Paul N., 1999. Global climate science, uncertainty and politics: Data-laden models, model-filtered data, in: *Science as Culture* 8, 437–472. https://doi.org/10.1080/09505439909526558

Epstein, Steven, 1995. The Construction of Lay Expertise: AIDS Activism and the Forging of Credibility in the Reform of Clinical Trials, in: *Science, Technology, & Human Values* 20, 408–437. https://doi.org/10.1177/016224399502000402

Evers, Marc, Stampf, Olaf, Traufetter, Gerald, 2010. Die Wolkenschieber, in: *Der Spiegel*, 29 March 2010, 140–149.

Eyring, Veronika, Cox, Peter M., Flato, Gregory M., Gleckler, Peter J., Abramowitz, Gab, Caldwell, Peter, Collins, William D., Gier, Bettina K., Hall, Alex D., Hoffman, Forrest M., Hurtt, George C., Jahn, Alexandra, Jones, Chris D., Klein, Stephen A., Krasting, John P., Kwiatkowski, Lester, Lorenz, Ruth, Maloney, Eric, Meehl, Gerald A., Pendergrass, Angeline G., Pincus, Robert, Ruane, Alex C., Russell, Joellen L., Sanderson, Benjamin M., Santer, Benjamin D., Sherwood, Steven C., Simpson, Isla R., Stouffer, Ronald J., Williamson, Mark S., 2019. Taking climate model evaluation to the next level, in: *Nature Climate Change* 9, 102–110. https://doi.org/10.1038/s41558-018-0355-y

Fantl, Jeremy, 2017. Knowledge How, in: Zalta, E.N. (Ed.), *The Stanford Encyclopedia of Philosophy*. Fall 2017 Edition, Metaphysics Research Lab, Stanford University, https://plato.stanford.edu/archives/fall2017/entries/knowledge-how/.

Feest, Uljana, Steinle, Friedrich, 2016. Experiments, in: Humphrey, P. (Ed.), *The Oxford Handbook of Philosophy of Science*. Oxford: Oxford University Press, pp. 274–295.

Feyerabend, Paul Karl, 1959. An Attempt at a Realistic Interpretation of Experience, in: Feyerabend, P.K. (Ed.), *Realism, Rationalism, and Scientific Method: Philosophical Papers*. Cambridge: Cambridge University Press, pp. 17–36.

Fidler, Fiona, Wilcox, John, 2021. Reproducibility of Scientific Results, in: Zalta, E.N. (Ed.), *The Stanford Encyclopedia of Philosophy*. Summer 2021 Edition, Metaphysics Research Lab, Stanford University, https://plato.stanford.edu/archives/sum2021/entries/scientific-reproducibility/.

Fine, Gary Alan, 2010. *Authors of the Storm: Meteorologists and the Culture of prediction*. Chicago; London: University of Chicago Press.

Flato, Gregory, Marotzke, Jochem, Abiodum, Babatunde, Braconnot, Pascale, Chou, Sin Chan, Collins, William, Cox, Peter, Driouech, Fatima, Emori, Seita, Eyring, Veronika, Forest, Chris, Gleckler, Peter J., Guilyardi, Eric, Jakob, Christian, Kattsov, Vladimir, Reason, Chris, Rummukaien, Markku, 2013. Evaluation of Climate Models, in: Stocker, T.F., Qin, D., Plattner, G.-K., Tignor, M., Allen, S.K., Boschung, J., Nauels, A., Xia, Y., Bex, V., Midgley, P.M. (Eds.), *Climate Change 2013: The Physical Science Basis. Contribution of Working Group I to the Fifth Assessment Report of the Intergovernmental Panel on Climate Change*. Cambridge: Cambridge University Press, pp. 741–866.

Fleck, Ludwig, [1935] 1979. *Genesis and Development of a Scientific Fact*. Bradley, F., Trenn, T. J. (Trans.), Chicago; London: University of Chicago Press.

Folland, Chris. K., Parker, David. E., 1995. Correction of instrumental biases in historical sea surface temperature data, in: *Quarterly Journal of the Royal Meteorological Society* 121, 319–367. https://doi.org/10.1002/qj.49712152206

Forster, Piers, Storelvmo, Trude, Armour, Kyle, Dufresne, Jean-Louis, Frame, David, Lunt, Daniel J., Mauritsen, Thorsten, Palmer, Matthew D., Watanabe, Masahiro, Wild, Martin, Zhang, Hua, 2021. The Earth's Energy Budget, Climate Feedbacks and Climate Sensitivity, in: Masson-Delmotte, V., Zhai, P., Pirani, A., Connors, S.L., Péan, C., Berger, S., Caud, N., Chen, Y., Goldfarb, L., Gomis, M.I., Huang, M., Leitzell, K., Lonnoy, E., Matthews, J.B.R., Maycock, T.K., Waterfield, T., Yelekçi, O., Yu, R., Zhou, B. (Eds.), *Climate Change 2021: The Physical Science Basis. Contribution of Working Group I*

to the Sixth Assessment Report of the Intergovernmental Panel on Climate Change. Cambridge, New York: Cambridge University Press, pp. 923–1054.

Frank, David, Esper, Jan, Zorita, Eduardo, Wilson, Rob, 2010. A noodle, hockey stick, and spaghetti plate: a perspective on high-resolution paleoclimatology, in: *Wiley Interdisciplinary Reviews: Climate Change* 1, 507–516. https://doi.org/10.1002/wcc.53

Frank, Philipp G., 1953. The variety of reasons for the acceptance of scientific theories, in: Frank, P.G. (Ed.), *The Validation of Scientific Theories*. New York: Collier Books, pp. 13–26.

Franssen, Maarten, Lokhorst, Gert-Jan, van de Poel, Ibo, 2018. Philosophy of Technology, Fall 2018 Edition, in: Zalta, E.N. (Ed.), *The Stanford Encyclopedia of Philosophy*. Metaphysics Research Lab, Stanford University, https://plato.stanford.edu/archives/fall2018/entries/technology/.

Frigg, Roman, Reiss, Julian, 2009. The philosophy of simulation: hot new issues or same old stew?, in: *Synthese* 169, 593–613. https://doi.org/10.1007/s11229-008-9438-z

Frisch, Mathias, 2015. Predictivism and old evidence: a critical look at climate model tuning, in: *European Journal for Philosophy of Science* 5, 171–190. https://doi.org/10.1007/s13194-015-0110-4

Gaffen, Dian J., Sargent, Michael A., Habermann, Ray E., Lanzante, John R., 2000. Sensitivity of Tropospheric and Stratospheric Temperature Trends to Radiosonde Data Quality, in: *Journal of Climate* 13, 1776–1796. https://doi.org/10.1175/1520-0442(2000)013<1776:SOTAST>2.0.CO;2

Giere, Ronald N., 2006. *Scientific perspectivism*. Chicago: University of Chicago Press.

Goldman, Alvin I., 2001. Experts: Which Ones Should You Trust?, in: *Philosophy and Phenomenological Research* 63, 85–110.

Gooding, David, 2000. Experiment, in: Newton-Smith, W. (Ed.), *A Companion to the Philosophy of Science*. Blackwell, Oxford, pp. 117–126.

Goodwin, William Mark, 2015. Global Climate Modeling as Applied Science, in: *Journal for General Philosophy of Science* 46, 339–350. https://doi.org/10.1007/s10838-015-9301-0

Gramelsberger, Gabriele, 2011. What do numerical (climate) models really represent?, in: *Studies in History and Philosophy of Science Part A* 42, 296–302. https://doi.org/10.1016/j.shpsa.2010.11.037

Guillemot, Hélène, 2017. How to develop climate models? The "gamble" of improving climate model parametrizations, in: Heymann, M., Gramels-

berger, G., Mahony, M. (Eds.), *Cultures of Prediction in Atmospheric and Climate Science*. New York: Routledge, pp. 120–136.

Guillemot, Hélène, 2010. Connections between simulations and observation in climate computer modeling. Scientist's practices and "bottom-up epistemology" lessons, in: *Studies in History and Philosophy of Science Part B: Studies in History and Philosophy of Modern Physics* 41, 242–252. https://doi.org/10.1016/j.shpsb.2010.07.003

Gundersen, Torbjørn, 2020. Value-Free yet Policy-Relevant? The Normative Views of Climate Scientists and Their Bearing on Philosophy, in: *Perspectives on Science* 28, 89–118. https://doi.org/10.1162/posc_a_00334

Hacking, Ian, 1983. Representing and Intervening: Introductory topics on the philosophy of natural science. Cambridge University Press, Cambridge.

Hall, Alex, Cox, Peter, Huntingford, Chris, Klein, Stephen, 2019. Progressing emergent constraints on future climate change, in: *Nature Climate Change* 9, 269–278. https://doi.org/10.1038/s41558-019-0436-6

Hall, Alex, Qu, Xin, 2006. Using the current seasonal cycle to constrain snow albedo feedback in future climate change, in: *Geophysical Research Letters* 33, L03502. https://doi.org/10.1029/2005GL025127

Hampshire, Stuart, 1949. Fallacies in Moral Philosophy, in: *Mind* 58, 466–482.

Hanea, Anca M., Hemming, Victoria, Nane, Gabriela F., 2021. Uncertainty Quantification with Experts: Present Status and Research Needs, in: *Risk Analysis* 42, 254–263. https://doi.org/10.1111/risa.13718

Hanson, Norwood Russell, 1958. *Patterns of Discovery. An Inquiry into the Conceptual Foundations of Science*. Cambridge: Cambridge University Press.

Haraway, Donna, 1989. *Primate Visions: Gender, Race, and Nature in the World of Modern Science*. New York: Routledge.

Harris, Adam J L, Corner, Adam, Xu, Juemin, Du, Xiufang, 2013. Lost in translation? Interpretations of the probability phrases used by the Intergovernmental Panel on Climate Change in China and the UK, in: *Climatic Change* 121, 415–425. https://doi.org/10.1007/s10584-013-0975-1

Harris, Todd, 2003. Data Models and the Acquisition and Manipulation of Data, in: *Philosophy of Science* 70, 1508–1517. https://doi.org/10.1086/377426

Hartmann, Dennis L., Klein-Tank, Albert M. G., Rusticucci, Matilde, Alexander, Lisa V., Brönnimann, Stefan, Charabi, Yassine Abdul-Rahman, Dentener, Frank J., Dlugokencky, Edward J., Easterling, David R., Kaplan, Alexey, Soden, Brian J., Thorne, Peter W., Wild, Martin, Zhai, Panmao, 2013. Observations: Atmosphere and Surface Supplementary Material, in:

Stocker, T.F., Qin, D., Tignor, M., Allen, S.K., Boschung, J., Nauels, A., Xia, Y., Bex, V., Midgley, P.M. (Eds.), *Climate Change 2013: The Physical Science Basis. Contribution of Working Group I to the Fifth Assessment Report of the Intergovernmental Panel on Climate Change*. Cambridge: Cambridge University Press, pp. 159–254.

Harvard, Stephanie, Winsberg, Eric, 2022. The Epistemic Risk in Representation, in: *Kennedy Institute of Ethics Journal* 32, 1–31. https://doi.org/10.1353/ken.2022.0001

Held, Isaac Meyer, 2005. The Gap between Simulation and Understanding in Climate Modeling, in: *Bulletin of the American Meteorological Society* 86, 1609–1614. https://doi.org/10.1175/BAMS-86-11-1609

Hempel, Carl, 1965. *Aspects of the scientific explanation*. New York: The Free Press.

Henning, Korinna, Drosten, Christian, 2020. Coronavirus-Update Folge 40, 12 December 2002, https://www.ndr.de/nachrichten/info/coronaskript194.pdf.

Heymann, Matthias, Gramelsberger, Gabriele, Mahony, Martin, 2017a. Key characteristics of cultures of prediction, in: Heymann, M., Gramelsberger, G., Mahony, M. (Eds.), *Cultures of Prediction in Atmospheric and Climate Science: Epistemic and Cultural Shifts in Computer-Based Modelling and Simulation*. London, New York: Routledge, pp. 18–41.

Heymann, Matthias, Gramelsberger, Gabriele, Mahony, Martin, 2017b. Introduction, in: Heymann, M., Gramelsberger, G., Mahony, M. (Eds.), *Cultures of Prediction in Atmospheric and Climate Science: Epistemic and Cultural Shifts in Computer-Based Modelling and Simulation*. London; New York: Routledge, pp. 1–17.

Hillerbrand, Rafaela, 2014. Climate Simulations: Uncertain Projections for an Uncertain World, in: *Journal for General Philosophy of Science* 45, 17–32. https://doi.org/10.1007/s10838-014-9266-4

Hillerbrand, Rafaela, 2010. On Non-Propositional Aspects in Modelling Complex Systems. in: *Analyse & Kritik* 01, 107–120.

Hourdin, Frédéric, Mauritsen, Thorsten, Gettelman, Andrew, Golaz, Jean-Christophe, Balaji, Venkatramani, Duan, Qingyun, Folini, Doris, Ji, Duoying, Klocke, Daniel, Qian, Yun, Rauser, Florian, Rio, Catherine, Tomassini, Lorenzo, Watanabe, Masahiro, Williamson, Daniel, 2017. The Art and Science of Climate Model Tuning, in: *Bulletin of the American Meteorological Society* 98, 589–602. https://doi.org/10.1175/BAMS-D-15-00135.1

Howard, Don, 2006. Lost Wanderers in the Forest of Knowledge: Some Thoughts on the Discovery-Justification Distinction, in: Schickore, J.,

Steinle, F. (Eds.), *Revisiting Discovery and Justification. Historical and Philosophical Perspectives on the Context Distinction*. Dordrecht: Springer, pp. 3–22.

Howe, Joshua P., 2014. Behind the Curve: Science and the Politics of Global Warming. Washington: University of Washington Press.

Hoyningen-Huene, Paul, 2006. Context of Discovery versus Context of Justification and Thomas Kuhn, in: Schickore, J., Steinle, F. (Eds.), *Revisiting Discovery and Justification. Historical and Philosophical Perspectives on the Context Distinction*. Dordrecht: Springer, pp. 119–131.

Huebner, Bryce, Kukla, Rebecca, Winsberg, Eric, 2017. Making an Author in Radically Collaborative Research, in: Boyer-Kassem, T., Mayo-Wilson, C., Weisberg, M. (Eds.), *Scientific Collaboration and Collective Knowledge*. Oxford: Oxford University Press, pp. 95–116.

Hume, David, [1739–1740] 1888. *A treatise of human nature*. Oxford: Clarendon Press.

Humphrey, Paul, 2004. *Extending ourselves. Computational science, empiricism, and scientific method*. New York: Oxford University Press.

Intemann, Kristen, 2017. Who Needs Consensus Anyway? Addressing Manufactured Doubt and Increasing Public Trust in Climate Science, in: *Public Affairs Quarterly* 31, 189–208.

Intemann, Kristen, 2015. Distinguishing between legitimate and illegitimate values in climate modeling, in: *European Journal for Philosophy of Science* 5, 217–232. https://doi.org/10.1007/s13194-014-0105-6

IPCC, 2023. Structure of the IPCC. https://www.ipcc.ch/about/structure/ (accessed 30 April 2023).

IPCC, 2021a. Annex VII: Glossary [J.B. Robin. Matthews, Vincent. Möller, Renée van Diemen, Jan S. Fuglestvedt, Valérie Masson-Delmotte, Carlos Méndez, Sergey Semenov, Andy Reisinger (Eds.)], in: Masson-Delmotte, V., Zhai, P., Pirani, A., Connors, S.L., Péan, C., Berger, S., Caud, N., Chen, Y., Goldfarb, L., Gomis, M.I., Huang, M., Leitzell, K., Lonnoy, E., Matthews, J.B.R., Maycock, T.K., Waterfield, T., Yelekçi, O., Yu, R., Zhou, B. (Eds.), *Climate Change 2021: The Physical Science Basis. Contribution of Working Group I to the Sixth Assessment Report of the Intergovernmental Panel on Climate Change*. Cambridge, New York: Cambridge University Press, pp. 3–32.

IPCC, 2021b. Summary for Policymakers, in: Masson-Delmotte, V., Zhai, P., Pirani, A., Connors, S.L., Péan, C., Berger, S., Caud, N., Chen, Y., Goldfarb, L., Gomis, M.I., Huang, M., Leitzell, K., Lonnoy, E., Matthews, J.B.R., Maycock, T.K., Waterfield, T., Yelekçi, O., Yu, R., Zhou, B. (Eds.), *Climate Change 2021: The Physical Science Basis. Contribution of Working Group I to the Sixth As-*

sessment Report of the Intergovernmental Panel on Climate Change. Cambridge, New York: Cambridge University Press, pp. 3–32.

Isaac, Alistair M.C., 2014. Model uncertainty and policy choice: A plea for integrated subjectivism, in: *Studies in History and Philosophy of Science Part A* 47, 42–50. https://doi.org/10.1016/j.shpsa.2014.05.004

Jebeile, Julie, Ardourel, Vincent, 2019. Verification and Validation of Simulations Against Holism, in: *Minds and Machines* 29, 149–168. https://doi.org/10.1007/s11023-019-09493-8

Jebeile, Julie, Crucifix, Michel, 2021. Value management and model pluralism in climate science, in: *Studies in History and Philosophy of Science* 88, 120–127. https://doi.org/10.1016/j.shpsa.2021.06.004

Jeffrey, Richard C., 1956. Valuation and Acceptance of Scientific Hypotheses, in: *Philosophy of Science* 23, 237–246.

Johnson, Ann, 2017. Philosophical and Conceptual Issues, in: Lenhard, J., Carrier, M. (Eds.), *Mathematics as a Tool. Tracing New Roles of Mathematics in the Sciences*, Boston Studies in the Philosophy and History of Science. Springer International Publishing, Cham, pp. 23–35.

Jones, Phil, 1999. Diagram for Wmo Statement. https://wikileaks.org/wiki/Climatic_Research_Unit_emails,_data,_models,_1996-2009/ (accessed 17 May 2023).

Keller, Evelyn Foxx, 1985. *Reflections on Gender and Science*. New Haven: Yale University Press.

Kennedy, John J., 2014. A review of uncertainty in in situ measurements and data sets of sea surface temperature, in: *Reviews of Geophysics* 52, 1–32. https://doi.org/10.1002/2013RG000434

Kent, Elizabeth C., Kennedy, John, Berry, David I., Smith, Robert O., 2010. Effects of instrumentation changes on sea surface temperature measured *in situ*, in: *Wiley Interdisciplinary Reviews: Climate Change* 1, 718–728. https://doi.org/10.1002/wcc.55

Kitcher, Philip, 2001. *Science, truth, and democracy*. Oxford: Oxford University Press.

Kitcher, Philip, 1993. *The Advancement of Science. Science without Legend, Objectivity without Illusions*. Oxford: Oxford University Press.

Knorr-Cetina, Karin, 1999. *Epistemic Cultures. How the Science make Knowledge*. Cambridge: Harvard University Press.

Knorr-Cetina, Karin, 1981. *The Manufacture of Knowledge: An Essay on the Constructivist and Contextual Nature of Science*. Oxford: Pergamon Press.

Knutti, Reto, 2018. Climate Model Confirmation: From Philosophy to Predicting Climate in the Real World, in: A. Lloyd, E., Winsberg, E. (Eds.), *Climate Modelling: Philosophical and Conceptual Issues*. Cham: Palgrave Macmillan, pp. 325–359. https://doi.org/10.1007/978-3-319-65058-6_11

Knutti, Reto, Hegerl, Gabriele C., 2008. The Equilibrium Sensitivity of the Earth's Temperature to Radiation Change, in: *Nature Geoscience* 1, 735–743.

Knutti, Reto, Masson, David, Gettelman, Andrew, 2013. Climate model genealogy: Generation CMIP5 and how we got there, in: *Geophysical Research Letters* 40, 1194–1199. https://doi.org/10.1002/grl.50256

Knutti, Reto, Sedláček, Jan, Sanderson, Benjamin M., Lorenz, Ruth, Fischer, Erich M., Eyring, Veronika, 2017. A climate model projection weighting scheme accounting for performance and interdependence, in: *Geophysical Research Letters* 44, 1909–1918. https://doi.org/10.1002/2016GL072012

Kosolosky, Laszlo, 2015. "Peer Review is Melting Our Glaciers": What Led the Intergovernmental Panel on Climate Change (IPCC) to Go Astray?, in: *Journal for General Philosophy of Science* 46, 351–366. https://doi.org/10.1007/s10838-015-9303-y

Kuhn, Thomas S, 1977. *The Essential Tension: Selected Studies in Scientific Tradition and Change*. Chicago: University of Chicago Press.

Kuhn, Thomas S., 1962. *The Structure of Scientific Revolutions*, Chicago: University of Chicago Press.

Kukla, Rebecca, 2012. "Author TBD": Radical Collaboration in Contemporary Biomedical Research, in: *Philosophy of Science* 79, 845–858. https://doi.org/10.1086/668042

Lam, Vincent, Majszak, Mason M., 2022. Climate tipping points and expert judgment, in: *WIREs Climate Change* 13. https://doi.org/10.1002/wcc.805

Landström, Catharine, 2017. Tracing uncertainty management through four IPCC Assessment Reports and beyond, in: Heymann, M., Gramelsberger, G., Mahony, M. (Eds.), *Cultures of Prediction in Atmospheric and Climate Science: Epistemic and Cultural Shifts in Computer-Based Modelling and Simulation*. London, New York: Routledge, pp. 214–230.

Latour, Bruno, Woolgar, Steve, 1979. *Laboratory Life: The Construction of Scientific Facts*. Los Angeles: Sage Publications.

Laudan, Larry, 2004. The Epistemic, the Cognitive, and the Social, in: Machamer, P., Wolters, G. (Eds.), *Science Values and Objectivity*. Pittsburgh: University of Pittsburgh Press, pp. 14–23. https://doi.org/10.2307/j.ctt5vkg7t

Lee, June-Yi, Marotzke, Jochem, Bala, Govindasamy, Cao, Long, Corti, Susanna, Dunne, John P., Engelbrecht, Francois, Fischer, Erich, Fyfe, John C., Mutemi, Joseph, Ndiaye, Ousmane, Panickal, Swapna, Zhou, Tianjun, 2021. Future Global Climate: Scenario-Based Projections and Near-Term Information, in: Masson-Delmotte, V., Zhai, P., Pirani, A., Connors, S.L., Péan, C., Berger, S., Caud, N., Chen, Y., Goldfarb, L., Gomis, M.I., Huang, M., Leitzell, K., Lonnoy, E., Matthews, J.B.R., Maycock, T.K., Waterfield, T., Yelekçi, O., Yu, R., Zhou, B. (Eds.), *Climate Change 2021: The Physical Science Basis. Contribution of Working Group I to the Sixth Assessment Report of the Intergovernmental Panel on Climate Change*. Cambridge, New York: Cambridge University Press, Cambridge, pp. 553–672.

Lehner, Flavio, Deser, Clara, Maher, Nicola, Marotzke, Jochem, Fischer, Erich M., Brunner, Lukas, Knutti, Reto, Hawkins, Ed, 2020. Partitioning climate projection uncertainty with multiple large ensembles and CMIP5/6, in: *Earth System Dynamics* 11, 491–508. https://doi.org/10.5194/esd-11-491-2020

Lenhard, Johannes, 2020. Understanding and Complexity – The Dilemma of Growth, in: Gramelsberger, G., Lenhard, J., Parker, W.S. (Eds.), Philosophical Perspectives on Earth System Modeling: Truth, Adequacy, and Understanding. *Journal of Advances in Modeling Earth Systems*, e2019MS001720.

Lenhard, Johannes, 2019. *Calculated surprises: a philosophy of computer simulation*. New York: Oxford University Press.

Lenhard, Johannes, 2018. Holism, or the Erosion of Modularity: A Methodological Challenge for Validation, in: *Philosophy of Science* 85, 832–844. https://doi.org/10.1086/699675

Lenhard, Johannes, Winsberg, Eric, 2010. Holism, entrenchment, and the future of climate model pluralism, in: *Studies in History and Philosophy of Science Part B: Studies in History and Philosophy of Modern Physics* 41, 253–262. https://doi.org/10.1016/j.shpsb.2010.07.001

Leonelli, Sabina, 2019. What distinguishes data from models?, in: *European Journal for Philosophy of Science* 9. https://doi.org/10.1007/s13194-018-0246-0

Leuschner, Anna, 2016. Die Unterschätzung des Klimawandels. Zum Einfluss nicht-epistemischer Werte auf die Klimamodellierung, in: Leuschner, A. (Ed.), *Die Energiewende Und Ihre Modelle. Was Uns Energieszenarien Sagen Können – Und Was Nicht*. Bielefeld: Transcript, pp. 75–88.

Leuschner, Anna, 2015. Uncertainties, Plurality, and Robustness in Climate Research and Modeling: On the Reliability of Climate Prognoses, in: *Journal for*

General Philosophy of Science 46, 367–381. https://doi.org/10.1007/s10838-015-9304-x

Leuschner, Anna, 2012a. Pluralism and objectivity: Exposing and breaking a circle, in: *Studies in History and Philosophy of Science Part A* 43, 191–198. https://doi.org/10.1016/j.shpsa.2011.12.030

Leuschner, Anna, 2012b. *Die Glaubwürdigkeit der Wissenschaft. Eine wissenschafts- und erkenntnistheoretische Analyse am Beispiel der Klimaforschung*. Bielefeld: Transcript.

Levi, Isaac, 1960. Must the Scientist Make Value Judgments?, in: *The Journal of Philosophy* 57, 345. https://doi.org/10.2307/2023504

Levins, Richard, 1993. A Response to Orzack and Sober: Formal Analysis and the Fluidity of Science, in: *The Quarterly Review of Biology* 68, 547–555. https://doi.org/10.1086/418302

Levins, Richard, 1966. The Strategy of Model Building in Population Biology, in: *American Scientist* 54, 421–431.

Lloyd, Elisabeth A., 2015. Model Robustness as a Confirmatory Virtue: The Case of Climate Science, in: *Studies in History and Philosophy of Science Part A* 49, 58–68.

Lloyd, Elisabeth A., 2012. The role of 'complex' empiricism in the debates about satellite data and climate models, in: *Studies in History and Philosophy of Science Part A* 43, 390–401. https://doi.org/10.1016/j.shpsa.2012.02.001

Lloyd, Elisabeth A., 2010. Confirmation and Robustness of Climate Models, in: *Philosophy of Science* 77, 971–984. https://doi.org/10.1086/657427

Lloyd, Elisabeth A., 2009. I – Elisabeth A. Lloyd: Varieties of Support and Confirmation of Climate Models, in: *Aristotelian Society Supplementary Volume* 83, 213–232. https://doi.org/10.1111/j.1467-8349.2009.00179.x

Løhre, Erik, Juanchich, Marie, Sirota, Miroslav, Teigen, Karl Halvor, Shepherd, Theodore G., 2019. Climate Scientists' Wide Prediction Intervals May Be More Likely but Are Perceived to Be Less Certain, in: *Weather, Climate, and Society* 11, 565–575. https://doi.org/10.1175/WCAS-D-18-0136.1

Longino, Helen, 2008. Values, Heuristics, and the Politics of Knowledge, in: Carrier, M., Howard, D., Kourany, J. (Eds.), *The Challenge of the Social and the Pressure of the Practical. Science and Values Revisited*. Pittsburgh: University of Pittsburgh Press, pp. 68–86.

Longino, Helen, 2002. *The fate of knowledge*. Princeton: Princeton University Press.

Longino, Helen, 1990. *Science as Social Knowledge. Values and Objectivity on Scientific Inquiry*. Princeton: Princeton University Press.

MacIntyre, Alasdair C., 1959. Hume on "Is" and "Ought," in: Chappell, V.C. (Ed.), *Hume: A Collection of Critical Essays*. New York: Doubleday.

Malik, Saira, 2017. Observation Versus Experiment: An Adequate Framework for Analysing Scientific Experimentation?, in: *Journal for General Philosophy of Science* 48, 71–95. https://doi.org/10.1007/s10838-016-9335-y

Mann, Michael E., 2018. Reconciling Climate Model/Data Discrepancy: The Case of the 'Tree That Didn't Bark, in: Lloyd, E.A., Winsberg, E. (Eds.), *Climate Modelling: Philosophical and Conceptual Issues*. Cham: Springer, pp. 175–197.

Mastrandrea, Michael D., Field, Christopher B., Stocker, Thomas F., Edenhofer, Ottmar, Ebi, Kristie, Frame, David J., Held, Hermann, Kriegler, Elmar, Mach, Katharine J., Matschoss, Patrick R., Plattner, Gian-Kasper, Yohe, Gary W., Zwiers, Francis W., 2010. *Guidance Note for Lead Authors of the IPCC Fifth Assessment Report on Consistent Treatment of Uncertainties*. Intergovernmental Panel on Climate Change (IPCC). https://www.ipcc.ch/site/assets/uploads/2017/08/AR5_Uncertainty_Guidance_Note.pdf

Mauritsen, Thorsten, Stevens, Bjorn, Roeckner, Erich, Crueger, Traute, Esch, Monika, Giorgetta, Marco, Haak, Helmuth, Jungclaus, Johann, Klocke, Daniel, Matei, Daniela, Mikolajewicz, Uwe, Notz, Dirk, Pincus, Robert, Schmidt, Hauke, Tomassini, Lorenzo, 2012. Tuning the climate of a global model, in: *Journal of Advances in Modeling Earth Systems* 4, M00A01. https://doi.org/10.1029/2012MS000154

McFarlane, Norman, 2011. Parameterizations: representing key processes in climate models without resolving them: Parameterizations, in: *Wiley Interdisciplinary Reviews: Climate Change* 2, 482–497. https://doi.org/10.1002/wcc.122

McMullin, Ernan, 1982. Values in Science, in: *PSA: Proceedings of the Biennial Meeting of the Philosophy of Science Association* 1982, 3–28.

Mears, Carl A., Schabel, Matthias C., Wentz, Frank J., 2003. A Reanalysis of the MSU Channel 2 Tropospheric Temperature Record, in: *Journal of Climate* 16, 3650–3664. https://doi.org/10.1175/1520-0442(2003)016<3650:AROTMC>2.0.CO;2

Medimorec, Srdan, Pennycook, Gordon, 2015. The language of denial: text analysis reveals differences in language use between climate change proponents and skeptics, in: *Climatic Change* 133, 597–605. https://doi.org/10.1007/s10584-015-1475-2

Megill, Alan, 1994. Introduction: Four Senses of Objectivity, in: Megill, A. (Ed.), *Rethinking Objectivity*. Durham: Duke University Press, pp. 1–20.

Merton, Robert K., 1973. *The Sociology of Science. Theoretical and Empirical Investigations*, Storer, N.W. (Ed.). Chicago: University of Chicago Press.

Mitchell, Sandra D., 2009. *Unsimple Truths: Science, Complexity and Policy*. Chicago: University of Chicago Press.

Morrison, Margaret, 2015. *Reconstructing Reality: Models, Mathematics, and Simulations*. Oxford: Oxford University Press. https://doi.org/10.1093/acprof:oso/9780199380275.001.0001

Nagel, Ernest, 1961. *The structure of science: Problems in the logic of scientific explanation*. New York: Brace & World.

Nickles, Thomas, 1980. Introductory essay: Scientific discovery and the future of philosophy of science, in: Nickles, T. (Ed.), *Scientific Discovery, Logic, and Rationality*, Boston Studies in the Philosophy of Science. Dordrecht: Reidel, pp. 1–59.

Nonaka, Ikujiro, Takeuchi, Hirotaka, 1995. *The Knowledge-creating Company: How Japanese Companies Create the Dynamics of Innovation*. New York: Oxford University Press.

Notz, Dirk, 2015. How well must climate models agree with observations?, in: *Philosophical Transactions of the Royal Society A: Mathematical, Physical and Engineering Sciences* 373, 20140164. https://doi.org/10.1098/rsta.2014.0164

Nunn, Patrick D., Reid, Nicholas J., 2016. Aboriginal Memories of Inundation of the Australian Coast Dating from More than 7000 Years Ago, in: *Australian Geographer* 47, 11–47. https://doi.org/10.1080/00049182.2015.1077539

Oppenheimer, Michael, Little, Christopher M., Cooke, Roger M., 2016. Expert judgement and uncertainty quantification for climate change, in: *Nature Climate Change* 6, 445–451. https://doi.org/10.1038/nclimate2959

Oreskes, Naomi, 2004. The Scientific Consensus on Climate Change, in: *Science* 306, 1686–1686. https://doi.org/10.1126/science.1103618

Oreskes, Naomi, Conway, Erik M., 2010. *Merchants of Doubt: How a Handful of Scientists Obscured the Truth on Issues from Tobacco Smoke to Global Warming*. New York: Bloomsbury Press.

Oreskes, Naomi, Shrader-Frechette, Kristin, Belitz, Kenneth, 1994. Verification, Validation, and Confirmation of Numerical Models in the Earth Sciences, in: Science 263, 641–646. https://doi.org/10.1126/science.263.5147.641

Orzack, Steven Hecht, Sober, Elliott, 1993. A Critical Assessment of Levins's The Strategy of Model Building in Population Biology (1966), in: *The Quarterly Review of Biology* 68, 533–546. https://doi.org/10.1086/418301

Oxburgh, Ronald, Davies, Huw, Emanuel, Kerry A., Graumlich, Lisa, Hand, David, Huppert, Herbert, Kelly, Michael, 2010. *Report of the International Panel set up by the University of East Anglia to examine the research of the Climatic Research Unit.* 14 April 2010, University of East Anglia.

Parker, Wendy S., 2020. Local Model-Data Symbiosis in Meteorology and Climate Science, in: *Philosophy of Science* 87, 807–818. https://doi.org/10.1086/710621

Parker, Wendy S., 2018. Climate Science, in: Zalta, E.N. (Ed.), *The Stanford Encyclopedia of Philosophy*. Summer 2018 Edition, Metaphysics Research Lab, Stanford University, https://plato.stanford.edu/archives/sum2018/entries/climate-science/.

Parker, Wendy S., 2017. Computer Simulation, Measurement, and Data Assimilation, in: *The British Journal for the Philosophy of Science* 68, 273–304. https://doi.org/10.1093/bjps/axv037

Parker, Wendy S., 2014. Values and uncertainties in climate prediction, revisited, in: *Studies in History and Philosophy of Science Part A* 46, 24–30. https://doi.org/10.1016/j.shpsa.2013.11.003

Parker, Wendy S., 2011. When Climate Models Agree: The Significance of Robust Model Predictions, in: *Philosophy of Science* 78, 579–600. https://doi.org/10.1086/661566

Parker, Wendy S., 2010. Predicting weather and climate: Uncertainty, ensembles and probability, in: *Studies in History and Philosophy of Science Part B: Studies in History and Philosophy of Modern Physics* 41, 263–272. https://doi.org/10.1016/j.shpsb.2010.07.006

Parker, Wendy S., 2009. II – Wendy S. Parker: Confirmation and adequacy-for-Purpose in Climate Modelling, in: *Aristotelian Society Supplementary Volume* 83, 233–249. https://doi.org/10.1111/j.1467-8349.2009.00180.x

Parker, Wendy S., Winsberg, Eric, 2018. Values and evidence: how models make a difference, in: *European Journal for Philosophy of Science* 8, 125–142. https://doi.org/10.1007/s13194-017-0180-6

Pew Research Center, 2019. Climate Change Still Seen as the Top Global Threat, but Cyberattacks a Rising Concern, February 2019. https://www.pewresearch.org/global/wp-content/uploads/sites/2/2019/02/Pew-Research-Center_Global-Threats-2018-Report_2019-02-10.pdf.

Pew Research Center, 2015. Public and Scientists' View on Science and Society, 29 January 2015, https://www.pewresearch.org/internet/wp-content/uploads/sites/9/2015/01/PI_ScienceandSociety_Report_012915.pdf.

Pickering, Andrew, 1992. From Science as Knowledge to Science as Practice, in: Pickering, A. (Ed.), *Science as Practice and Culture*. Chicago: University of Chicago Press, pp. 1–26.

Polanyi, Michael, [1958] 1962. *Personal knowledge. Towards a post-critical philosophy.* Chicago: University of Chicago Press.

Polanyi, Michael, 1966a. *The tacit dimension*. Chicago: University of Chicago Press.

Polanyi, Michael, 1966b. The Logic of Tacit Inference, in: *Philosophy* 41, 1–18. https://doi.org/10.1017/S0031819100066110

Popper, Karl, [1935] 1959. *The Logic of Scientific Discovery*. London; New York: Routledge.

Proctor, Robert N., 1991. *Values-Free Science? Purity and Power in Modern Knowledge*. Cambridge: Harvard University Press.

Quine, Willard Van Orman, 1951. Two Dogmas of Empiricism, in: *From a Logical Point of View*. Cambridge: Harvard University Press, pp. 20–46.

Reichenbach, Hans, 1938. *Experience and Prediction: An Analysis of the Foundations and Structure of Knowledge*. Chicago: University of Chicago Press.

Reiss, Julian, Sprenger, Jan, 2017. Scientific Objectivity, in: Zalta, E.N. (Ed.), *The Stanford Encyclopedia of Philosophy*. Winter 2017 Edition, Metaphysics Research Lab, Stanford University, https://plato.stanford.edu/archives/win2017/entries/scientific-objectivity/.

Rescher, Nicholas, 1988. *Predicting the future: An Introduction to the Theory of Forecasting*. New York: State University of New York Press.

Richardson, Lewis Fry, 1922. *Weather Prediction by Numerical Process*. Cambridge: Cambridge University Press.

Rooney, Phyllis, 1992. On Values in Science: Is the Epistemic/Non-Epistemic Distinction Useful?, in: *PSA: Proceedings of the Biennial Meeting of the Philosophy of Science Association* 1992, 13–22. https://doi.org/10.1086/psaprocbienmeetp.1992.1.192740

Rudner, Richard, 1953. The Scientist Qua Scientist Makes Value Judgments, in: *Philosophy of Science* 20, 1–6.

Ruphy, Stéphanie, 2016. *Scientific Pluralism Reconsidered. A New Approach to the (Dis)Unity of Science*. Pittsburgh: University of Pittsburgh Press, Pittsburgh.

Rusbridger, Alan, 2015. Climate change: why the Guardian is putting threat to Earth front and centre, in: *the Guardian*, 6 March 2015, https://www.theguardian.com/environment/2015/mar/06/climate-change-guardian-threat-to-earth-alan-rusbridger (accessed 17 May 2023).

Ryle, Gilbert, [1949] 1973. *The concept of mind*. Harmondsworth; Ringwood: Penguin University Books.

Santer, Benjamin D, Fyfe, John C, Pallotta, Giuliana, Flato, Gregory M, Meehl, Gerald A, England, Matthew H, Hawkins, Ed, Mann, Michael E, Painter, Jeffrey F, Bonfils, Céline, Cvijanovic, Ivana, Mears, Carl, Wentz, Frank J, Po-Chedley, Stephen, Fu, Qiang, Zou, Cheng-Zhi, 2017. Causes of differences in model and satellite tropospheric warming rates, in: *Nature Geoscience* 10, 478–485. https://doi.org/10.1038/ngeo2973

Santer, Benjamin D., Hnilo, Justin J., Wigley, Tom M. L., Boyle, James S., Doutriaux, Charles, Fiorino, Michael, Parker, David E., Taylor, Karl E., 1999. Uncertainties in observationally based estimates of temperature change in the free atmosphere, in: *Journal of Geophysical Research: Atmospheres* 104, 6305–6333. https://doi.org/10.1029/1998JD200096

Santer, Benjamin David, Thorne, Peter W., Haimberger, Leopold, Taylor, Karl E., Wigley, Tom M. L., Lanzante, John R., Solomon, Susan, Free, Melissa, Gleckler, Peter J., Jones, Philip Douglas, Karl, Thomas R., Klein, Sidnei A., Mears, Carl, Nychka, Douglas, Schmidt, Gavin A., Sherwood, Steven C., Wentz, Frank J., 2008. Consistency of modelled and observed temperature trends in the tropical troposphere, in: *International Journal of Climatology* 28, 1703–1722. https://doi.org/10.1002/joc.1756

Schickore, Jutta, Steinle Friedrich, 2006a. Introduction: Revisiting Discovery and Justification, in: Schickore, J., Steinle F. (Eds.), Revisiting Discovery and Justification. Historical and philosophical perspectives on the context distinction. Dordrecht: Springer, pp. vii–xix. https://doi.org/10.1007/1-4020-4251-5_1

Schickore, Jutta, Steinle, Friedrich, 2006b. Revisiting Discovery and Justification. Historical and philosophical perspectives on the context distinction. Dordrecht: Springer.

Schmidt, Gavin A., 2007. Past reconstructions: problems, pitfalls and progress. realclimate.org. http://www.realclimate.org/index.php/archives/2007/12/past-reconstructions/ (accessed 17 May 2023).

Schmidt, Gavin A., Bader, David, Donner, Leo J., Elsaesser, Gregory S., Golaz, Jean-Christophe, Hannay, Cecile, Molod, Andrea, Neale, Richard B., Saha, Suranjana, 2017. Practice and philosophy of climate model tuning across six US modeling centers, in: *Geoscientific Model Development* 10, 3207–3223. https://doi.org/10.5194/gmd-10-3207-2017

Schmidt, Gavin A., Sherwood, Steven C., 2015. A practical philosophy of complex climate modelling, in: *European Journal for Philosophy of Science* 5, 149–169. https://doi.org/10.1007/s13194-014-0102-9

Schupbach, Jonah N., 2018. Robustness Analysis as Explanatory Reasoning, in: *The British Journal for the Philosophy of Science* 69, 275–300. https://doi.org/10.1093/bjps/axw008

Searle, John R., 1980. Minds, brains, and programs, in: *Behavioral and Brain Sciences* 3, 417–424. https://doi.org/10.1017/S0140525X00005756

Shapere, Dudley, 1982. The Concept of Observation in Science and Philosophy, in: *Philosophy of Science* 49, 485–525

Sherwood, S. C., Webb, M. J., Annan, J. D., Armour, K. C., Forster, P. M., Hargreaves, J. C., Hegerl, G., Klein, S. A., Marvel, K. D., Rohling, E. J., Watanabe, M., Andrews, T., Braconnot, P., Bretherton, C. S., Foster, G. L., Hausfather, Z., Heydt, A. S., Knutti, R., Mauritsen, T., Norris, J. R., Proistosescu, C., Rugenstein, M., Schmidt, G. A., Tokarska, K. B., Zelinka, M. D., 2020. An Assessment of Earth's Climate Sensitivity Using Multiple Lines of Evidence, in: *Reviews of Geophysics* 58, e2019RG000678. https://doi.org/10.1029/2019RG000678

Shewhart, Walter A., Demming, William E., 1939. *Statistical Methods from the Viewpoint of Quality Control.* North Chelmsford: Courier Corporation.

Skelton, Maurice, Porter, James J., Dessai, Suraje, Bresch, David N., Knutti, Reto, 2017. The social and scientific values that shape national climate scenarios: a comparison of the Netherlands, Switzerland and the UK, in: *Regional Environmental Change* 17, 2325–2338. https://doi.org/10.1007/s10113-017-1155-z

Slivinski, Laura C., Compo, Gilbert P., Whitaker, Jeffrey S., Sardeshmukh, Prashant D., Giese, Benjamin S., McColl, Chesley, Allan, Rob, Yin, Xungang, Vose, Russell S., Titchner, Holly, Kennedy, John, Spencer, Lawrence J., Ashcroft, Linden, Brönnimann, Stefan, Brunet, Manola, Camuffo, Dario, Cornes, Richard, Cram, Thomas A., Crouthamel, Richard, Domínguez-Castro, Fernando, Freeman, J. Eric, Gergis, Joëlle, Hawkins, Ed, Jones, Philip Douglas, Jourdain, Sylvie, Kaplan, Alexey, Kubota, Hisayuki, Le Blancq, Frank, Lee, Tsz-Cheung, Lorrey, Andrew, Luterbacher, Jürg, Maugeri, Maurizio, Mock, Cary J., Moore, G. W. Kent, Przybylak, Rajmund, Pudmenzky, Christa, Reason, Chris, Slonosky, Victoria C., Smith, Catherine A., Tinz, Birger, Trewin, Blair, Valente, Maria Antónia, Wang, Xiaolan L., Wilkinson, Clive, Wood, Kevin, Wyszyński, Przemysław, 2019. Towards a more reliable historical reanalysis: Improvements for version 3 of the

Twentieth Century Reanalysis system, in: *Quarterly Journal of the Royal Meteorological Society* 145, 2876–2908. https://doi.org/10.1002/qj.3598

Soler, Léna, 2011. Tacit aspects of experimental practices: analytical tools and epistemological consequences, in: *European Journal for Philosophy of Science* 1, 393–433. https://doi.org/10.1007/s13194-011-0039-1

Soler, Léna, Zwart, Sjoerd D., 2013. Editorial Introduction: Collins and Tacit Knowledge, in: *Philosophia Scientiae* 5–23. https://doi.org/10.4000/philosophiascientiae.880

Spencer, Roy W., Christy, John R., 1990. Precise Monitoring of Global Temperature Trends from Satellites, in: *Science* 247, 1558–1562. https://doi.org/10.1126/science.247.4950.1558

Stainforth, David Alan, Aina, Tolu, Christensen, Carl, Collins, Matthew, Faull, Nicholas, Frame, David J., Kettleborough, Jamie A., Knight, Susie, Martin, Andrew, Murphy, James M., Piani, Claudio, Sexton, David, Smith, Leonard A., Spicer, Robert A., Thorpe, Alan J., Allen, Myles R., 2005. Uncertainty in predictions of the climate response to rising levels of greenhouse gases, in: *Nature* 433, 403–406. https://doi.org/10.1038/nature03301

Stanford, Kyle, 2023. Underdetermination of Scientific Theory, in: Zalta, E.N., Nodelman, U. (Eds.), *The Stanford Encyclopedia of Philosophy*. Summer 2023 Edition, Metaphysics Research Lab, Stanford University, https://plato.stanford.edu/archives/sum2023/entries/scientific-underdetermination/.

Steel, Daniel, 2015. Acceptance, values, and probability, in: *Studies in History and Philosophy of Science Part A* 53, 81–88. https://doi.org/10.1016/j.shpsa.2015.05.010

Steele, Katie, Werndl, Charlotte, 2013. Climate Models, Calibration, and Confirmation, in: *The British Journal for the Philosophy of Science* 64, 609–635.

Steinle, Friedrich, 2006. Concept Formation and the Limits of Justification: "Discovering" the two Electricities, in: Schickore, J., Steinle, F. (Eds.), *Revisiting Discovery and Justification. Historical and philosophical perspectives on the context distinction*. Dordrecht: Springer, pp. 183–195. https://doi.org/10.1007/1-4020-4251-5_11

Stevens, Bjorn, Sherwood, Steven C., Bony, Sandrine, Webb, Mark J., 2016. Prospects for narrowing bounds on Earth's equilibrium climate sensitivity, in: *Earth's Future* 4, 512–522. https://doi.org/10.1002/2016EF000376

Strevens, Michael, 2016. Complexity Theory, in: Humphrey, P. (Ed.), *The Oxford Handbook of Philosophy of Science*. Oxford: Oxford University Press, pp. 695–716.

Sundberg, Mikaela, 2011. The dynamics of coordinated comparisons: How simulationists in astrophysics, oceanography and meteorology create standards for results, in: *Social Studies of Science* 41, 107–125. https://doi.org/10.1177/0306312710385743

Suppes, Patrick, 1962. Models of Data, in: Nagel, E., Suppes, P., Tarski, A. (Eds.), *Logic, Methodology, and Philosophy of Science: Proceedings of the 1960 International Congress*. Stanford: Stanford University Press, pp. 252–261.

Supran, Geoffrey, Oreskes, Naomi, 2017. Assessing ExxonMobil's climate change communications (1977–2014), in: *Environmental Research Letters* 12, 084019. https://doi.org/10.1088/1748-9326/aa815f

Tebaldi, Claudia, Knutti, Reto, 2007. The use of the multi-model ensemble in probabilistic climate projections, in: *Philosophical Transactions of the Royal Society A: Mathematical, Physical and Engineering Sciences* 365, 2053–2075. https://doi.org/10.1098/rsta.2007.2076

Thompson, David W. J., Kennedy, John J., Wallace, John M., Jones, Phil D., 2008. A large discontinuity in the mid-twentieth century in observed global-mean surface temperature, in: *Nature* 453, 646–649. https://doi.org/10.1038/nature06982

Thompson, Erica, Frigg, Roman, Helgeson, Casey, 2016. Expert Judgment for Climate Change Adaptation, in: *Philosophy of Science* 83, 1110–1121. https://doi.org/10.1086/687942

Thorne, Peter W., Lanzante, John R., Peterson, Thomas C., Seidel, Dian J., Shine, Keith P., 2011. Tropospheric temperature trends: history of an ongoing controversy, in: *Wiley Interdisciplinary Reviews: Climate Change* 2, 66–88. https://doi.org/10.1002/wcc.80

van Fraassen, Bas C., 2008. *Scientific Representation*. Oxford: Oxford University Press.

van Fraassen, Bas C., 1980. *The scientific image*. Oxford: Oxford University Press.

Van Lange, Paul A. M., Joireman, Jeff, Milinski, Manfred, 2018. Climate Change: What Psychology Can Offer in Terms of Insights and Solutions, in: *Current Directions in Psychological Science* 27, 269–274. https://doi.org/10.1177/0963721417753945

Vinnikov, Konstantin Y., Grody, Norman C., 2003. Global Warming Trend of Mean Tropospheric Temperature Observed by Satellites, in: *Science* 302, 269–272. https://doi.org/10.1126/science.1087910

Washington, Haydn, Cook, John, 2011. *Climate Change Denial Heads in the Sand*. Oxfordshire: Earthscan.

Weart, Spencer, 2010. The development of general circulation models of climate, in: *Studies in History and Philosophy of Science Part B: Studies in History and Philosophy of Modern Physics* 41, 208–217. https://doi.org/10.1016/j.shpsb.2010.06.002

Weaver, Warren, 1948. Science and Complexity, in: *American Scientist* 36, 449–456. https://doi.org/10.1007/978-1-4899-0718-9_30

Weber, Max, 1946. *Essays in Sociology*, Gerth, H.H., Mills, C.W. (Trans.). New York: Oxford University Press.

Weber, Max, 1913. Über einige Kategorien der verstehenden Soziologie, in: Winckelmann, J. (Ed.), *Gesammelte Aufsätze Zur Wissenschaftslehre*. Mohr Siebeck, Tübingen.

Weisberg, Michael, 2006. Robustness Analysis, in: *Philosophy of Science* 73, 730–742.

Wentz, Frank J., Schabel, Matthias, 2000. Precise climate monitoring using complementary satellite data sets, in: *Nature* 403, 414–416. https://doi.org/10.1038/35000184

Werndl, Charlotte, 2019. Initial-Condition Dependence and Initial-Condition Uncertainty in Climate Science, in: *The British Journal for the Philosophy of Science* 70, 953–976. https://doi.org/10.1093/bjps/axy021

Wilholt, Torsten, 2009. Bias and Values in Scientific Research, in: *Studies in History and Philosophy of Science* 40, 92–101. https://doi.org/10.1016/j.shpsa.2008.12.005

Wimsatt, William, 2007. *Re-engineering philosophy for limited beings. Piecewise approximations to reality*. Cambridge: Harvard University Press.

Winsberg, Eric, 2019. Computer Simulations in Science, in: Zalta, E.N. (Ed.), *The Stanford Encyclopedia of Philosophy*. Winter 2019 Edition, Metaphysics Research Lab, Stanford University, https://plato.stanford.edu/archives/win2019/entries/simulations-science/.

Winsberg, Eric, 2018. *Philosophy and Climate Science, Philosophy and Climate Science*. Cambridge: Cambridge University Press. https://doi.org/10.1017/9781108164290

Winsberg, Eric, 2012. Values and Uncertainties in the Predictions of Global Climate Models, in: *Kennedy Institute of Ethics Journal* 22, 111–137. https://doi.org/10.1353/ken.2012.0008

Winsberg, Eric, 2010. *Science in the age of computer simulation*. Chicago: University of Chicago Press.

Winsberg, Eric, 2003. Simulated Experiments: Methodology for a Virtual World, in: *Philosophy of Science* 70, 105–125. https://doi.org/10.1086/367872

Winsberg, Eric, Huebner, Bryce, Kukla, Rebecca, 2014. Accountability and values in radically collaborative research, in: *Studies in History and Philosophy of Science Part A* 46, 16–23. https://doi.org/10.1016/j.shpsa.2013.11.007

Wittgenstein, Ludwig, 1953. *Philosophical Investigations*, Anscombe, G.E.M. (Tran.). Oxford: Blackwell.

Zängl, Günther, Reinert, Daniel, Rípodas, Pilar, Baldauf, Michael, 2015. The ICON (ICOsahedral Non-hydrostatic) modelling framework of DWD and MPI-M: Description of the non-hydrostatic dynamical core, in: *Quarterly Journal of the Royal Meteorological Society* 141, 563–579. https://doi.org/10.1002/qj.2378

Zickfeld, Kirsten, Levermann, Anders, Morgan, M. Granger, Kuhlbrodt, Till, Rahmstorf, Stefan, Keith, David W., 2007. Expert judgements on the response of the Atlantic meridional overturning circulation to climate change, in: *Climatic Change* 82, 235–265. https://doi.org/10.1007/s10584-007-9246-3